Advancing Maths for AQA

Revise for CORE 2

Tony Clough

Series editors
Sam Boardman   Roger Williamson   Ted Graham

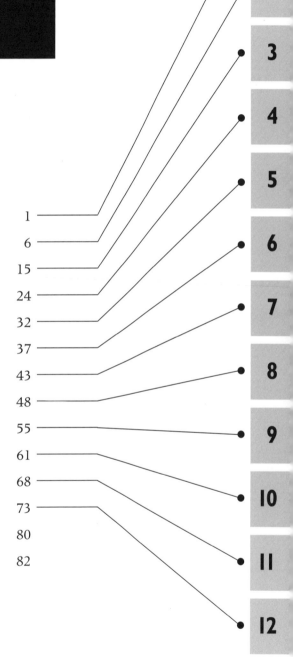

heinemann.co.uk
✓ Free online support
✓ Useful weblinks
✓ 24 hour online ordering

01865 888058

Heinemann
Inspiring generations

D0453024

Heinemann is an imprint of Pearson Education Limited, a company incorporated in England and Wales, having its registered office at Edinburgh Gate, Harlow, Essex, CM20 2JE. Registered company number: 872828

www.heinemann.co.uk

Heinemann is a registered trademark of
Pearson Education Limited

© Tony Clough 2005
Complete work © Heinemann Educational Publishers 2005

First published 2005

11
10 9 8 7 6 5

British Library Cataloguing in Publication Data is available from the British Library on request.

ISBN: 978 0 435 513573

Typeset and illustrated by Tech-Set Limited, Gateshead, Tyne & Wear
Original illustrations © Harcourt Education Limited, 2005
Cover design by mccdesign ltd
Printed in Malaysia, CTP-KHL

About this book

This book is designed to help you get your best possible grade in your Pure Core Maths 2 examination. The author is a Principal examiner, and has a good understanding of AQA's requirements.

Revise for Core 2 covers the key topics that are tested in the Core 2 exam paper. You can use this book to help you revise at the end of your course, or you can use it throughout your course alongside the course textbook, *Advancing Maths for AQA AS & A level Pure Core Maths 1 & 2*, which provides complete coverage of the syllabus.

Helping you prepare for your exam

To help you prepare, each topic offers you:

- **Key points to remember** – summarise the mathematical ideas you need to know and be able to use.

- **Worked examples** – help you understand and remember important methods, and show you how to set out your answers clearly.

- **Revision exercises** – help you practise using these important methods to solve problems. Exam-level questions are included so you can be sure you are reaching the right standard, and answers are given at the back of the book so you can assess your progress.

- **Test Yourself questions** – help you see where you need extra revision and practice. If you do need extra help, they show you where to look in the *Advancing Maths for AQA AS & A level Pure Core Maths 1 & 2* textbook and which example to refer to in this book.

Exam practice and advice on revising

Examination style paper – this paper at the end of the book provides a set of questions of examination standard. It gives you an opportunity to practise taking a complete exam before you meet the real thing. The answers are given at the back of the book.

How to revise – for advice on revising before the exam, read the *How to revise* section on the next page.

How to revise using this book

Making the best use of your revision time

The topics in this book have been arranged in a logical sequence so you can work your way through them from beginning to end. But **how** you work on them depends on how much time there is between now and your examination.

If you have plenty of time before the exam then you can **work through each topic in turn**, covering the key points and worked examples before doing the revision exercises and Test Yourself questions.

If you are short of time then you can **work through the Test Yourself sections** first, to help you see which topics you need to do further work on.

However much time you have to revise, make sure you break your revision into short blocks of about 40 minutes, separated by five- or ten-minute breaks. Nobody can study effectively for hours without a break.

Using the Test Yourself sections

Each Test Yourself section provides a set of key questions. Try each question:

- If you can do it and get the correct answer, then move on to the next topic. Come back to this topic later to consolidate your knowledge and understanding by working through the key points, worked examples and revision exercises.

- If you cannot do the question, or get an incorrect answer or part answer, then work through the key points, worked examples and revision exercises before trying the Test Yourself questions again. If you need more help, the cross-references beside each Test Yourself question show you where to find relevant information in the *Advancing Maths for AQA AS & A level Pure Core Maths 1 & 2* textbook and which example in *Revise for C2* to refer to.

Reviewing the key points

Most of the key points are straightforward ideas that you can learn: try to understand each one. Imagine explaining each idea to a friend in your own words, and say it out loud as you do so. This is a better way of making the ideas stick than just reading them silently from the page.

As you work through the book, remember to go back over key points from earlier topics at least once a week. This will help you to remember them in the exam.

CHAPTER 1

Indices

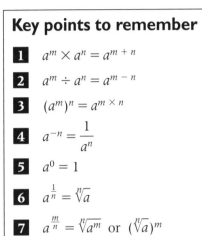

Key points to remember

1 $a^m \times a^n = a^{m+n}$

2 $a^m \div a^n = a^{m-n}$

3 $(a^m)^n = a^{m \times n}$

4 $a^{-n} = \dfrac{1}{a^n}$

5 $a^0 = 1$

6 $a^{\frac{1}{n}} = \sqrt[n]{a}$

7 $a^{\frac{m}{n}} = \sqrt[n]{a^m}$ or $(\sqrt[n]{a})^m$

Worked example 1

Simplify $x^3 y^2 \times (2x^2 y^4)^3$.

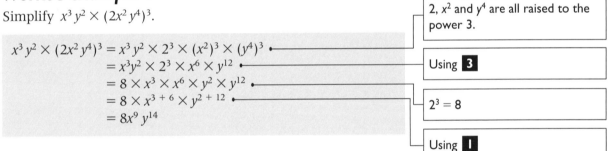

$$
\begin{aligned}
x^3 y^2 \times (2x^2 y^4)^3 &= x^3 y^2 \times 2^3 \times (x^2)^3 \times (y^4)^3 \\
&= x^3 y^2 \times 2^3 \times x^6 \times y^{12} \\
&= 8 \times x^3 \times x^6 \times y^2 \times y^{12} \\
&= 8 \times x^{3+6} \times y^{2+12} \\
&= 8x^9 y^{14}
\end{aligned}
$$

2, x^2 and y^4 are all raised to the power 3.

Using **3**

$2^3 = 8$

Using **1**

Worked example 2

Simplify $\dfrac{(4m^2 n^4)^3}{8(mn^2)^5}$.

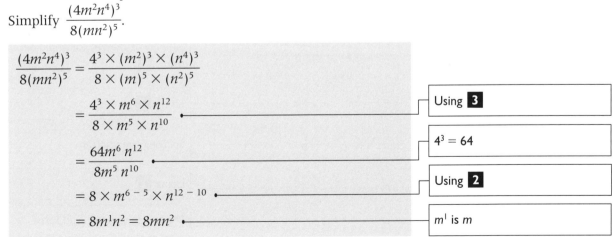

$$
\begin{aligned}
\frac{(4m^2 n^4)^3}{8(mn^2)^5} &= \frac{4^3 \times (m^2)^3 \times (n^4)^3}{8 \times (m)^5 \times (n^2)^5} \\[2mm]
&= \frac{4^3 \times m^6 \times n^{12}}{8 \times m^5 \times n^{10}} \\[2mm]
&= \frac{64 m^6 n^{12}}{8 m^5 n^{10}} \\[2mm]
&= 8 \times m^{6-5} \times n^{12-10} \\[2mm]
&= 8m^1 n^2 = 8mn^2
\end{aligned}
$$

Using **3**

$4^3 = 64$

Using **2**

m^1 is m

Worked example 3

Simplify $32(x^{-2})^{-4} \times (-4y^{-3}x^4)^{-2}$.

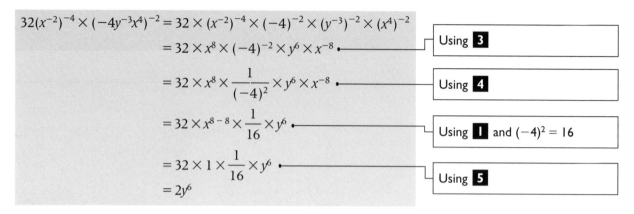

$$32(x^{-2})^{-4} \times (-4y^{-3}x^4)^{-2} = 32 \times (x^{-2})^{-4} \times (-4)^{-2} \times (y^{-3})^{-2} \times (x^4)^{-2}$$

$$= 32 \times x^8 \times (-4)^{-2} \times y^6 \times x^{-8}$$ — Using **3**

$$= 32 \times x^8 \times \frac{1}{(-4)^2} \times y^6 \times x^{-8}$$ — Using **4**

$$= 32 \times x^{8-8} \times \frac{1}{16} \times y^6$$ — Using **1** and $(-4)^2 = 16$

$$= 32 \times 1 \times \frac{1}{16} \times y^6$$ — Using **5**

$$= 2y^6$$

Worked example 4

Write $\sqrt{p} \div \sqrt[5]{p^2}$ as p raised to a single power.

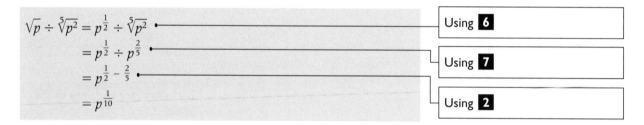

$$\sqrt{p} \div \sqrt[5]{p^2} = p^{\frac{1}{2}} \div \sqrt[5]{p^2}$$ — Using **6**

$$= p^{\frac{1}{2}} \div p^{\frac{2}{5}}$$ — Using **7**

$$= p^{\frac{1}{2} - \frac{2}{5}}$$ — Using **2**

$$= p^{\frac{1}{10}}$$

Worked example 5

Write $\dfrac{\left(x^{\frac{1}{4}} + x^{\frac{1}{2}}\right)\left(x^{\frac{1}{4}} - x^{\frac{1}{2}}\right)}{\sqrt[3]{x^2}}$ in the form $x^a - x^b$.

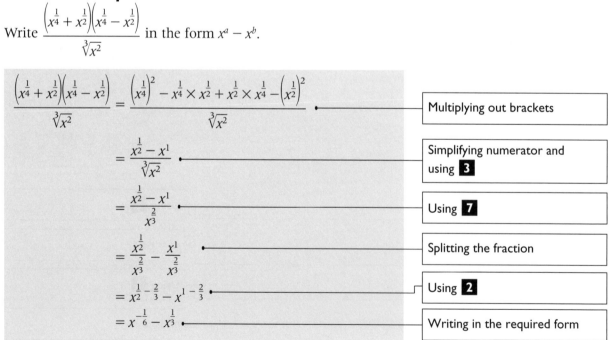

$$\frac{\left(x^{\frac{1}{4}} + x^{\frac{1}{2}}\right)\left(x^{\frac{1}{4}} - x^{\frac{1}{2}}\right)}{\sqrt[3]{x^2}} = \frac{\left(x^{\frac{1}{4}}\right)^2 - x^{\frac{1}{4}} \times x^{\frac{1}{2}} + x^{\frac{1}{2}} \times x^{\frac{1}{4}} - \left(x^{\frac{1}{2}}\right)^2}{\sqrt[3]{x^2}}$$ — Multiplying out brackets

$$= \frac{x^{\frac{1}{2}} - x^1}{\sqrt[3]{x^2}}$$ — Simplifying numerator and using **3**

$$= \frac{x^{\frac{1}{2}} - x^1}{x^{\frac{2}{3}}}$$ — Using **7**

$$= \frac{x^{\frac{1}{2}}}{x^{\frac{2}{3}}} - \frac{x^1}{x^{\frac{2}{3}}}$$ — Splitting the fraction

$$= x^{\frac{1}{2} - \frac{2}{3}} - x^{1 - \frac{2}{3}}$$ — Using **2**

$$= x^{-\frac{1}{6}} - x^{\frac{1}{3}}$$ — Writing in the required form

Worked example 6

Solve the equation $2^x \times 4^{2x+3} = 8\sqrt{2}$.

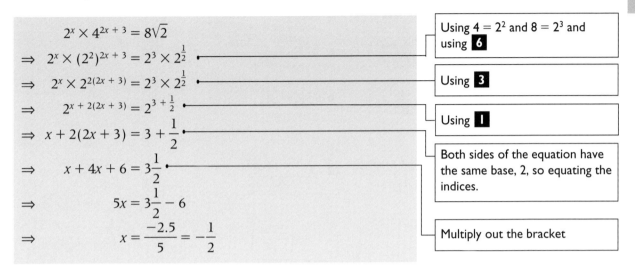

$$2^x \times 4^{2x+3} = 8\sqrt{2}$$

$$\Rightarrow\ 2^x \times (2^2)^{2x+3} = 2^3 \times 2^{\frac{1}{2}}$$

$$\Rightarrow\ 2^x \times 2^{2(2x+3)} = 2^3 \times 2^{\frac{1}{2}}$$

$$\Rightarrow\ 2^{x+2(2x+3)} = 2^{3+\frac{1}{2}}$$

$$\Rightarrow\ x + 2(2x+3) = 3 + \frac{1}{2}$$

$$\Rightarrow\ x + 4x + 6 = 3\frac{1}{2}$$

$$\Rightarrow\ 5x = 3\frac{1}{2} - 6$$

$$\Rightarrow\ x = \frac{-2.5}{5} = -\frac{1}{2}$$

Using $4 = 2^2$ and $8 = 2^3$ and using **6**

Using **3**

Using **1**

Both sides of the equation have the same base, 2, so equating the indices.

Multiply out the bracket

Worked example 7

Use the substitution $y = 9^x$ to solve the equation
$3 \times 81^x - 4 \times 9^x + 1 = 0$.

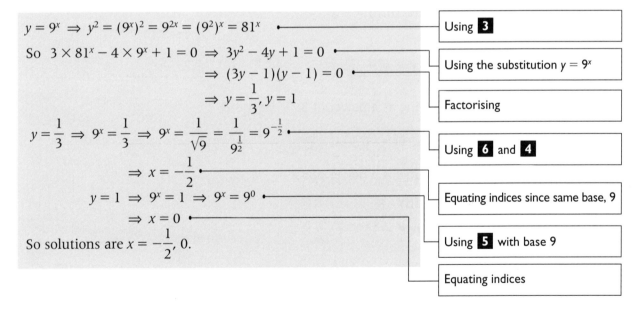

$$y = 9^x \Rightarrow y^2 = (9^x)^2 = 9^{2x} = (9^2)^x = 81^x$$

$$\text{So } 3 \times 81^x - 4 \times 9^x + 1 = 0 \Rightarrow 3y^2 - 4y + 1 = 0$$

$$\Rightarrow (3y-1)(y-1) = 0$$

$$\Rightarrow y = \frac{1}{3},\ y = 1$$

$$y = \frac{1}{3} \Rightarrow 9^x = \frac{1}{3} \Rightarrow 9^x = \frac{1}{\sqrt{9}} = \frac{1}{9^{\frac{1}{2}}} = 9^{-\frac{1}{2}}$$

$$\Rightarrow x = -\frac{1}{2}$$

$$y = 1 \Rightarrow 9^x = 1 \Rightarrow 9^x = 9^0$$

$$\Rightarrow x = 0$$

$$\text{So solutions are } x = -\frac{1}{2},\ 0.$$

Using **3**

Using the substitution $y = 9^x$

Factorising

Using **6** and **4**

Equating indices since same base, 9

Using **5** with base 9

Equating indices

REVISION EXERCISE 1

In questions 1–15 simplify as far as possible.

1 $2x^3 \times 3x^5$

2 $56a^7 \div 14a^3$

3 $(2n^2)^3$

4 $(m^3)^4 + (2m^6)^2$

5 $\dfrac{3p^2q^3 \times (2p^2q)^3}{4(pq)^4}$

6 $\dfrac{(3a^2b^3)^2 \times 4ab^2}{12(ab^2)^4}$

7 $\dfrac{(3m^2)^0 \times 4(m^3n^4)^2}{2m^{-2}n^3}$

8 $a^{-5} \div \dfrac{1}{24(-x^{-2})^{-4}a^3}$

9 $(2x)^{-3} \times$

10 $\dfrac{(m^3n^{-2})^{-1}}{\left(\dfrac{m}{n^2}\right)^3}$

11 $\dfrac{6x^2 \times 2\sqrt{x}}{(27x^3)^{\frac{1}{3}}}$

12 $\dfrac{2x^2 \times (16x)^{\frac{3}{4}}}{4\sqrt{x}}$

13 $\dfrac{\sqrt[4]{x^4y^2} \times \sqrt{4x}}{\sqrt[3]{8y}}$

14 $(\sqrt{2p})^5 \times \sqrt{8p^3q} \div \sqrt[3]{q}$

15 $\dfrac{\left(\dfrac{27\sqrt{y}}{x^2}\right)^{\frac{2}{3}}}{\sqrt[6]{\dfrac{64y}{x^5}}}$

16 Write $\dfrac{(\sqrt{x} + 1)(\sqrt{x} - 1)}{\sqrt[3]{x}}$ in the form $x^a - x^b$.

17 Write $\dfrac{(x\sqrt{x} - 1)(x\sqrt{x} + 1)}{\sqrt{x^3}}$ in the form $x^p - x^q$.

18 Write $\dfrac{(x\sqrt{x} + 1)^2}{\sqrt{x}}$ in the form $x^a + x^b + cx$.

19 (a) Write each of the following as a power of 3:
   (i) $\sqrt{3}$ **(ii)** 27^x
   (b) Hence solve the equation $27^x \times 3^{x+1} = \sqrt{3}$.

20 (a) Write each of the following as a power of 4:
   (i) 2 **(ii)** $\dfrac{1}{16}$ **(iii)** 8^x
   (b) Hence solve the equation $8^x \times 4^{2-x} = \dfrac{1}{16}$.

21 Solve the equation $\dfrac{9^x}{3^{1-3x}} = 3\sqrt{3}$.

22 (a) Show that the substitution $y = 4^x$ transforms the equation $16^x - 3 \times 4^x - 4 = 0$ into the quadratic equation $y^2 - 3y - 4 = 0$.
   (b) Hence show that the equation $16^x - 3 \times 4^x - 4 = 0$ has only one real solution and find its value.

23 Use the substitution $y = 9^x$ to solve the equation $9^{2x+1} - 28 \times 9^x + 3 = 0$.

Test yourself	**What to review**
	If your answer is incorrect:

1 Simplify each of the following:
  (a) $2p \times p^6 \times 3p^2$ **(b)** $m^2n \times 4nm^3 \times 2nk$
  (c) $18x^6y^5 \div 6yx^4$ **(d)** $(2a^3bc^4)^3$
  (e) $\dfrac{2(w^2z^3)^2 \times (2wz^2)^3}{4w(z^2)^2}.$
	See p 1 Examples 1 and 2 or review Advancing Maths for AQA C1C2 pp 208–211.

2 Simplify each of the following: **(a)** $(2p)^{-2} \times \dfrac{8}{p^2}$ **(b)** $(x^2y^3)^{-1} \div \left(\dfrac{y}{x}\right)^2$ **(c)** $\dfrac{8\left(\dfrac{2a^2b}{c}\right)^{-2}}{ab^{-1}c^2}.$	See p 2 Example 3 or review Advancing Maths for AQA C1C2 pp 211–214.
3 Write each of the following as a power of 16: **(a)** 4 **(b)** 2 **(c)** $\dfrac{1}{64}$ **(d)** $\sqrt{8}.$	Review Advancing Maths for AQA C1C2 pp 214–217.
4 Simplify $\dfrac{\sqrt[4]{x^2y^3} \times (xy^3)^{\frac{1}{2}}}{\sqrt{y}}.$	See p 2 Examples 4 and 5 or review Advancing Maths for AQA C1C2 pp 214–217.
5 Solve $\dfrac{4^x}{\sqrt[3]{4}} = 2.$	See p 3 Example 6 or review Advancing Maths for AQA C1C2 pp 217–219.
6 (a) Write 4^x as a power of 2. **(b)** Solve $(\sqrt{2})^{3x+2} \times \dfrac{1}{4^x} = 32.$	See p 3 Example 7 or review Advancing Maths for AQA C1C2 pp 217–219.

Test yourself ANSWERS

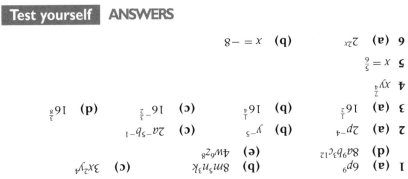

1 (a) $6p^9$
  (b) $8m^5n^3k$
  (c) $3x^2y^4$
  (d) $8a^9b^3c^{12}$
  (e) $4w^6z^8$

2 (a) $2p^{-4}$
  (b) y^{-5}
  (c) $2a^{-5}b^{-1}$

3 (a) $16^{\frac{1}{2}}$
  (b) $16^{\frac{1}{4}}$
  (c) $16^{-\frac{3}{2}}$
  (d) $16^{\frac{3}{8}}$

4 $xy^{\frac{7}{4}}$

5 $x = \frac{5}{6}$

6 (a) 2^{2x}
  (b) $x = -8$

CHAPTER 2

Further differentiation

Key points to remember

1 When differentiating, you can use the following strategy even when n is any rational number:

(a) if $y = x^n$, then $\dfrac{dy}{dx} = nx^{n-1}$,

(b) if $y = cx^n$ (c is a constant), then $\dfrac{dy}{dx} = cnx^{n-1}$,

(c) if $y = f(x) \pm g(x)$, then $\dfrac{dy}{dx} = f'(x) \pm g'(x)$.

2 The equation of the tangent to a curve at the point $P(x_1, y_1)$ is given by $y - y_1 = m(x - x_1)$, where m is the gradient of the curve at P.

3 The normal to a curve at the point P is a straight line that is perpendicular to the tangent at P. When the gradient of the tangent is m, the gradient of the normal is $-\dfrac{1}{m}$.

4 The nature of stationary points can be established using the second derivative test:

If P is a point where $\dfrac{dy}{dx} = 0$ and $\dfrac{d^2y}{dx^2}$ is negative then P is a maximum point.

If Q is a point where $\dfrac{dy}{dx} = 0$ and $\dfrac{d^2y}{dx^2}$ is positive then Q is a minimum point.

Worked example 1

Differentiate $y = \sqrt{x^3} - \dfrac{1}{2x}$ with respect to x.

$$y = \sqrt{x^3} - \frac{1}{2x} \Rightarrow y = (x^3)^{\frac{1}{2}} - \frac{1}{2} \times \frac{1}{x}$$

Using $\sqrt{x} = x^{\frac{1}{2}}$

$$\Rightarrow y = x^{\frac{3}{2}} - \frac{1}{2}x^{-1}$$

Using $(x^m)^n = x^{m \times n}$ and $\dfrac{1}{x^n} = x^{-n}$

$$\Rightarrow \frac{dy}{dx} = \frac{3}{2}x^{\frac{3}{2}-1} - \frac{1}{2}(-1x^{-1-1}) = \frac{3}{2}x^{\frac{1}{2}} + \frac{1}{2}x^{-2}$$

Differentiating using **1**

$$\Rightarrow \frac{dy}{dx} = \frac{3}{2}\sqrt{x} + \frac{1}{2x^2}$$

Using $x^{\frac{1}{2}} = \sqrt{x}$ and $x^{-n} = \dfrac{1}{x^n}$

2

Worked example 2

It is given that $f(x) = \sqrt{x}\left(x^3 - \dfrac{1}{x}\right)$. Find $f'(x)$.

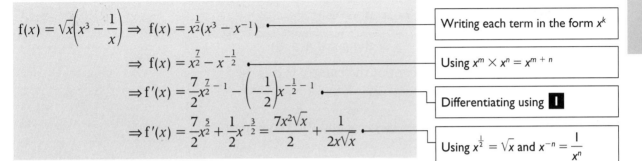

$$f(x) = \sqrt{x}\left(x^3 - \frac{1}{x}\right) \Rightarrow f(x) = x^{\frac{1}{2}}(x^3 - x^{-1})$$

Writing each term in the form x^k

$$\Rightarrow f(x) = x^{\frac{7}{2}} - x^{-\frac{1}{2}}$$

Using $x^m \times x^n = x^{m+n}$

$$\Rightarrow f'(x) = \frac{7}{2}x^{\frac{7}{2}-1} - \left(-\frac{1}{2}\right)x^{-\frac{1}{2}-1}$$

Differentiating using **I**

$$\Rightarrow f'(x) = \frac{7}{2}x^{\frac{5}{2}} + \frac{1}{2}x^{-\frac{3}{2}} = \frac{7x^2\sqrt{x}}{2} + \frac{1}{2x\sqrt{x}}$$

Using $x^{\frac{1}{2}} = \sqrt{x}$ and $x^{-n} = \dfrac{1}{x^n}$

Worked example 3

Differentiate $y = \dfrac{(2x+1)(x+4)}{2x^2}$ with respect to x.

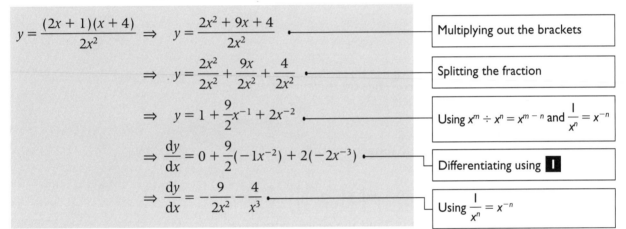

$$y = \frac{(2x+1)(x+4)}{2x^2} \Rightarrow y = \frac{2x^2 + 9x + 4}{2x^2}$$

Multiplying out the brackets

$$\Rightarrow y = \frac{2x^2}{2x^2} + \frac{9x}{2x^2} + \frac{4}{2x^2}$$

Splitting the fraction

$$\Rightarrow y = 1 + \frac{9}{2}x^{-1} + 2x^{-2}$$

Using $x^m \div x^n = x^{m-n}$ and $\dfrac{1}{x^n} = x^{-n}$

$$\Rightarrow \frac{dy}{dx} = 0 + \frac{9}{2}(-1x^{-2}) + 2(-2x^{-3})$$

Differentiating using **I**

$$\Rightarrow \frac{dy}{dx} = -\frac{9}{2x^2} - \frac{4}{x^3}$$

Using $\dfrac{1}{x^n} = x^{-n}$

Worked example 4

The diagram shows the curves $y = 4x^{\frac{1}{2}}$ and $y = x^{\frac{3}{2}}$ which intersect at the origin O and at the point A.

(a) Find the coordinates of A.

(b) Show that, at the point A, the gradient of one of the curves is three times the gradient of the other.

(c) Find the value of x for which the two curves have the same gradient.

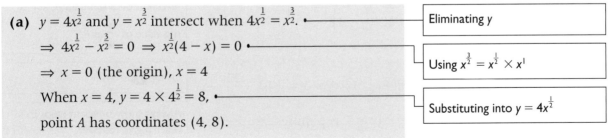

(a) $y = 4x^{\frac{1}{2}}$ and $y = x^{\frac{3}{2}}$ intersect when $4x^{\frac{1}{2}} = x^{\frac{3}{2}}$.

Eliminating y

$$\Rightarrow 4x^{\frac{1}{2}} - x^{\frac{3}{2}} = 0 \Rightarrow x^{\frac{1}{2}}(4-x) = 0$$

Using $x^{\frac{3}{2}} = x^{\frac{1}{2}} \times x^1$

$$\Rightarrow x = 0 \text{ (the origin)}, x = 4$$

When $x = 4$, $y = 4 \times 4^{\frac{1}{2}} = 8$,

Substituting into $y = 4x^{\frac{1}{2}}$

point A has coordinates $(4, 8)$.

(b) The gradient of the curve $y = 4x^{\frac{1}{2}}$ is given by

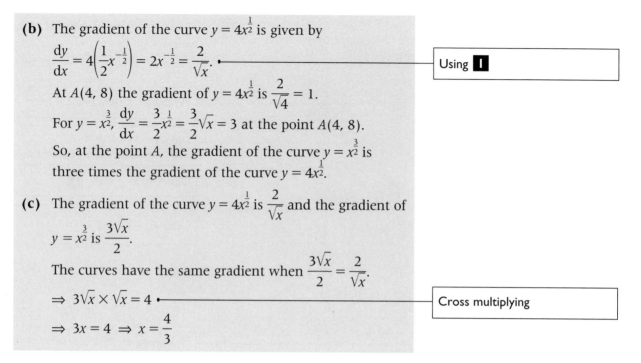

$$\frac{dy}{dx} = 4\left(\frac{1}{2}x^{-\frac{1}{2}}\right) = 2x^{-\frac{1}{2}} = \frac{2}{\sqrt{x}}.$$

Using **1**

At $A(4, 8)$ the gradient of $y = 4x^{\frac{1}{2}}$ is $\frac{2}{\sqrt{4}} = 1$.

For $y = x^{\frac{3}{2}}$, $\frac{dy}{dx} = \frac{3}{2}x^{\frac{1}{2}} = \frac{3}{2}\sqrt{x} = 3$ at the point $A(4, 8)$.

So, at the point A, the gradient of the curve $y = x^{\frac{3}{2}}$ is three times the gradient of the curve $y = 4x^{\frac{1}{2}}$.

(c) The gradient of the curve $y = 4x^{\frac{1}{2}}$ is $\frac{2}{\sqrt{x}}$ and the gradient of

$y = x^{\frac{3}{2}}$ is $\frac{3\sqrt{x}}{2}$.

The curves have the same gradient when $\frac{3\sqrt{x}}{2} = \frac{2}{\sqrt{x}}$.

$\Rightarrow 3\sqrt{x} \times \sqrt{x} = 4$

Cross multiplying

$\Rightarrow 3x = 4 \Rightarrow x = \frac{4}{3}$

Worked example 5

The point P, whose x-coordinate is $\frac{1}{2}$, lies on the curve C whose

equation is $y = \frac{x + 1}{x}$.

(a) Show that the equation of the tangent to the curve at P is
$y + 4x = 5$.

(b) The normal to the curve C at the point $\left(-2, \frac{1}{2}\right)$ intersects
the tangent at P at the point Q. Find the coordinates of Q.

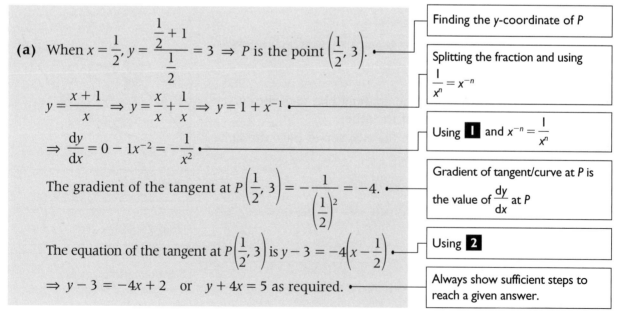

(a) When $x = \frac{1}{2}$, $y = \dfrac{\frac{1}{2} + 1}{\frac{1}{2}} = 3 \Rightarrow P$ is the point $\left(\frac{1}{2}, 3\right)$.

Finding the y-coordinate of P

$y = \frac{x + 1}{x} \Rightarrow y = \frac{x}{x} + \frac{1}{x} \Rightarrow y = 1 + x^{-1}$

Splitting the fraction and using $\frac{1}{x^n} = x^{-n}$

$\Rightarrow \frac{dy}{dx} = 0 - 1x^{-2} = -\frac{1}{x^2}$

Using **1** and $x^{-n} = \frac{1}{x^n}$

The gradient of the tangent at $P\left(\frac{1}{2}, 3\right) = -\dfrac{1}{\left(\frac{1}{2}\right)^2} = -4$.

Gradient of tangent/curve at P is the value of $\frac{dy}{dx}$ at P

The equation of the tangent at $P\left(\frac{1}{2}, 3\right)$ is $y - 3 = -4\left(x - \frac{1}{2}\right)$.

Using **2**

$\Rightarrow y - 3 = -4x + 2$ or $y + 4x = 5$ as required.

Always show sufficient steps to reach a given answer.

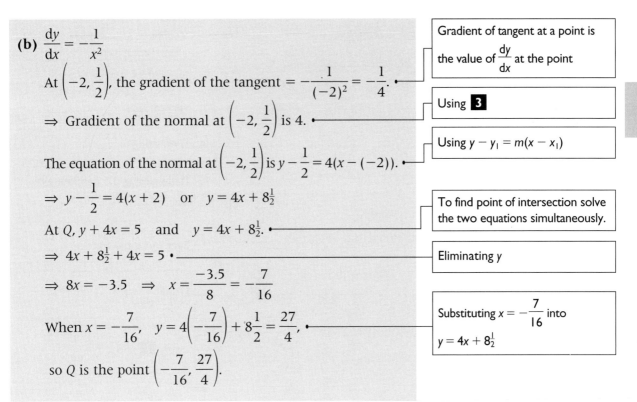

(b) $\dfrac{dy}{dx} = -\dfrac{1}{x^2}$

At $\left(-2, \dfrac{1}{2}\right)$, the gradient of the tangent $= -\dfrac{1}{(-2)^2} = -\dfrac{1}{4}$.

> Gradient of tangent at a point is the value of $\dfrac{dy}{dx}$ at the point

$\Rightarrow$ Gradient of the normal at $\left(-2, \dfrac{1}{2}\right)$ is 4.

> Using **3**

The equation of the normal at $\left(-2, \dfrac{1}{2}\right)$ is $y - \dfrac{1}{2} = 4(x - (-2))$.

> Using $y - y_1 = m(x - x_1)$

$\Rightarrow y - \dfrac{1}{2} = 4(x + 2)$ or $y = 4x + 8\frac{1}{2}$

At Q, $y + 4x = 5$ and $y = 4x + 8\frac{1}{2}$.

> To find point of intersection solve the two equations simultaneously.

$\Rightarrow 4x + 8\frac{1}{2} + 4x = 5$

> Eliminating y

$\Rightarrow 8x = -3.5 \Rightarrow x = \dfrac{-3.5}{8} = -\dfrac{7}{16}$

When $x = -\dfrac{7}{16}$, $y = 4\left(-\dfrac{7}{16}\right) + 8\dfrac{1}{2} = \dfrac{27}{4}$,

> Substituting $x = -\dfrac{7}{16}$ into $y = 4x + 8\frac{1}{2}$

so Q is the point $\left(-\dfrac{7}{16}, \dfrac{27}{4}\right)$.

Worked example 6

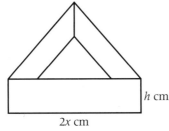

h cm

$2x$ cm

The diagram shows an open-topped metal container. The container consists of a thin base in the shape of an equilateral triangle with sides of length $2x$ cm, and three thin rectangular vertical faces each of width $2x$ cm and height h cm. The total internal surface area of the container is A cm². The container holds 729 cm³ of liquid when full.

(a) Given that the area of the triangular base is $\sqrt{3}x^2$ cm², show that $A = \sqrt{3}x^2 + \dfrac{1458\sqrt{3}}{x}$.

(b) Calculate the value of x for which A has a stationary value.

(c) By finding $\dfrac{d^2A}{dx^2}$, show that the stationary value is a minimum.

(d) Find the height of the container when its total internal surface area is a minimum. Give your answer in the form $k\sqrt{3}$ cm, where k is an integer.

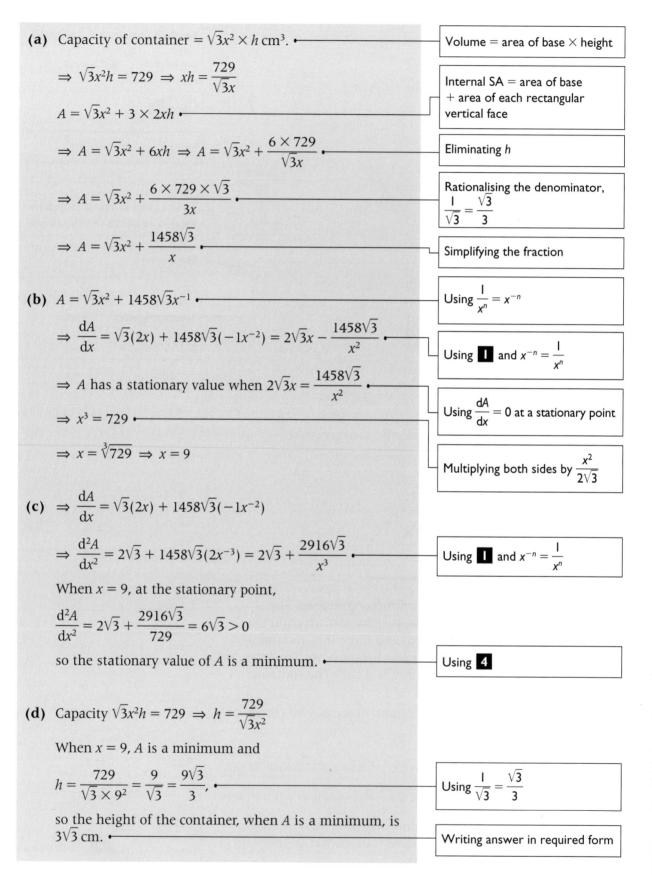

(a) Capacity of container $= \sqrt{3}x^2 \times h \text{ cm}^3$.

Volume = area of base × height

$$\Rightarrow \sqrt{3}x^2h = 729 \Rightarrow xh = \frac{729}{\sqrt{3}x}$$

Internal SA = area of base + area of each rectangular vertical face

$$A = \sqrt{3}x^2 + 3 \times 2xh$$

$$\Rightarrow A = \sqrt{3}x^2 + 6xh \Rightarrow A = \sqrt{3}x^2 + \frac{6 \times 729}{\sqrt{3}x}$$

Eliminating h

$$\Rightarrow A = \sqrt{3}x^2 + \frac{6 \times 729 \times \sqrt{3}}{3x}$$

Rationalising the denominator, $\frac{1}{\sqrt{3}} = \frac{\sqrt{3}}{3}$

$$\Rightarrow A = \sqrt{3}x^2 + \frac{1458\sqrt{3}}{x}$$

Simplifying the fraction

(b) $A = \sqrt{3}x^2 + 1458\sqrt{3}x^{-1}$

Using $\frac{1}{x^n} = x^{-n}$

$$\Rightarrow \frac{dA}{dx} = \sqrt{3}(2x) + 1458\sqrt{3}(-1x^{-2}) = 2\sqrt{3}x - \frac{1458\sqrt{3}}{x^2}$$

Using **1** and $x^{-n} = \frac{1}{x^n}$

$$\Rightarrow A \text{ has a stationary value when } 2\sqrt{3}x = \frac{1458\sqrt{3}}{x^2}$$

Using $\frac{dA}{dx} = 0$ at a stationary point

$$\Rightarrow x^3 = 729$$

$$\Rightarrow x = \sqrt[3]{729} \Rightarrow x = 9$$

Multiplying both sides by $\frac{x^2}{2\sqrt{3}}$

(c) $\Rightarrow \frac{dA}{dx} = \sqrt{3}(2x) + 1458\sqrt{3}(-1x^{-2})$

$$\Rightarrow \frac{d^2A}{dx^2} = 2\sqrt{3} + 1458\sqrt{3}(2x^{-3}) = 2\sqrt{3} + \frac{2916\sqrt{3}}{x^3}$$

Using **1** and $x^{-n} = \frac{1}{x^n}$

When $x = 9$, at the stationary point,

$$\frac{d^2A}{dx^2} = 2\sqrt{3} + \frac{2916\sqrt{3}}{729} = 6\sqrt{3} > 0$$

so the stationary value of A is a minimum.

Using **4**

(d) Capacity $\sqrt{3}x^2h = 729 \Rightarrow h = \frac{729}{\sqrt{3}x^2}$

When $x = 9$, A is a minimum and

$$h = \frac{729}{\sqrt{3} \times 9^2} = \frac{9}{\sqrt{3}} = \frac{9\sqrt{3}}{3},$$

Using $\frac{1}{\sqrt{3}} = \frac{\sqrt{3}}{3}$

so the height of the container, when A is a minimum, is $3\sqrt{3}$ cm.

Writing answer in required form

REVISION EXERCISE 2

In questions 1–18 find $\dfrac{dy}{dx}$.

2

1 $y = \dfrac{1}{x}$

2 $y = \sqrt{x}$

3 $y = \dfrac{1}{\sqrt{x}}$

4 $y = 1 + \dfrac{5}{x}$

5 $y = x + \dfrac{1}{2x^2}$

6 $y = 2x^2 - \dfrac{3}{2x^2} - 5$

7 $y = 4x + 6x^{\frac{1}{3}} - 1$

8 $y = 4x^{\frac{1}{2}} + \dfrac{2}{x^{\frac{1}{2}}} - 3x$

9 $y = \dfrac{5x^{\frac{2}{5}}}{6} + \dfrac{3}{2x^{\frac{1}{3}}} - 4$

10 $y = \dfrac{2}{3}\sqrt{x} - x\sqrt{x}$

11 $y = \sqrt[4]{16x^3} + 4x$

12 $y = \dfrac{\sqrt{x^3}}{6} - x^3\sqrt{x} + 2$

13 $y = 1 + \dfrac{5 + x}{x^2}$

14 $y = (\sqrt{x^3} + 4)^2$

15 $y = \dfrac{(1 + x)(1 - x)}{x^3}$

16 $y = \dfrac{(2 - 3x)(x - 2)}{2x^4}$

17 $y = \dfrac{6 + 4\sqrt{x} - \sqrt[3]{x}}{2x}$

18 $y = \dfrac{(2\sqrt{x} + 1)^2}{\sqrt[3]{x^2}}$

19 A curve has equation $y = 2x - \dfrac{1}{x^2}$.

    (a) Calculate the gradient of the curve at the point $\left(\dfrac{1}{2}, -3\right)$.

    (b) Find the value of x at which the gradient of the curve is $2\dfrac{1}{4}$.

20 A curve is defined for $x > 0$ by the equation
$y = \dfrac{(x - 1)(x - 2)}{x^3}$. Find the gradient of the curve at each of
the points where the curve crosses the x-axis.

21 Two curves are defined for $x > 0$ by the equations $y = \dfrac{x\sqrt{x}}{6} + 4$
and $y = 7 - \dfrac{1}{\sqrt{x}}$. Find, in surd form, the value of x for which
the two curves have the same gradient.

22 A curve is defined for $x > 0$ by the equation $y = 1 + \dfrac{2}{x} - 4\sqrt{x}$.

    (a) Find the equation of the tangent to the curve at the point
$(1, -1)$, giving your answer in the form $px + qy = r$, where p,
q and r are positive integers.

    (b) The tangent cuts the x-axis at the point A and the y-axis at
the point B. Find the area of triangle AOB, where O is the
origin $(0, 0)$.

23 A curve is defined for $x > 0$ by the equation $y = \dfrac{2\sqrt{x}(x + 3)}{x} - 5$.

(a) Find an equation of the tangent to the curve at the point $(1, 3)$.

(b) Show that the equation of the normal to the curve at the point where $x = 4$ is $y + 8x = 34$.

(c) The tangent at $(1, 3)$ intersects this normal at the point A. Find the x-coordinate of the point A.

24 The curve with equation $y = \sqrt{x}(x - 2)$ is defined for $x \geqslant 0$ and is shown below.

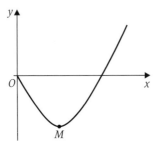

The curve has a minimum point M and O is the origin $(0, 0)$.

(a) Determine the x-coordinate of M.

(b) Find the equation of the normal to the curve at the point $P(1, -1)$, giving your answer in the form $ax + by = c$, where a, b and c are positive integers.

(c) The normal at P intersects the x-axis at the point A and the y-axis at the point B. Find the area of triangle AOB.

25 A curve is defined for $x > 0$ by the equation $y = f(x)$, where $f(x) = x\sqrt{x}(15 - x)$.

(a) Find the coordinates of the stationary point of the curve.

(b) Find the value of $\dfrac{d^2y}{dx^2}$ at the stationary point and hence determine its nature.

(c) Find the values of x for which f is a decreasing function.

26 A rectangular garden consists of a rectangular lawn of area $200\,\text{m}^2$ surrounded on three sides by a path of width $1\,\text{m}$. The width of the lawn is $x\,\text{m}$ as shown in the diagram. The area of the garden is $A\,\text{m}^2$.

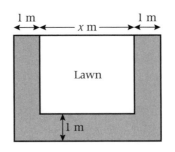

(a) Show that $A = x + \dfrac{400}{x} + 202$.

(b) Find the perimeter of the garden when its area is minimized.

Test yourself	**What to review**
	If your answer is incorrect:
1 Differentiate with respect to x: **(a)** $y = 3x - \dfrac{4}{x} + \dfrac{1}{2x^2}$ **(b)** $y = 9\sqrt[3]{x} + 4x^2\sqrt{x} + 3.$	See p 6 Example 1 or review Advancing Maths for AQA C1C2 pp 221–224.
2 Find f$'(x)$ for each of the following: **(a)** $f(x) = 2x(\sqrt{x} - x)^2$ **(b)** $f(x) = \dfrac{x\sqrt{x} + 2}{x^2}.$	See p 7 Examples 2 and 3 or review Advancing Maths for AQA C1C2 pp 225–226.

3 A curve has equation $y = \dfrac{2 + x}{\sqrt{x}}$ and P is the point on the curve where $x = 4$.
  (a) Find the gradient of the curve at P.
  (b) Find the equation of the tangent to the curve at P, giving your answer in the form $ax + by + c = 0$, where a, b and c are integers.
	See p 7 Examples 4 and 5 or review Advancing Maths for AQA C1C2 pp 227–228.

4 A curve has equation $y = \left(2x + \dfrac{1}{x}\right)^2$.
  (a) Show that the gradient of the curve at the point P, where $x = \dfrac{1}{2}$, is -12.
  (b) Find the equation of the normal to the curve at P, giving your answer in the form $ax + by + c = 0$, where a, b and c are integers.
	See p 8 Example 5 or review Advancing Maths for AQA C1C2 pp 227–228.

5 A curve has equation $y = \dfrac{4x^2 + 1}{x}$, $x \neq 0$.
  (a) Find $\dfrac{d^2y}{dx^2}$.
  (b) Show that the curve has two stationary points.
  (c) Find the coordinates of the maximum point of the curve.
	See p 9 Example 6 or review Advancing Maths for AQA C1C2 pp 229–232.

6 The width of a solid cuboid is x cm. The length of the cuboid is $3x$ cm and its height is h cm.
 	
The volume of the cuboid is 2304 cm³ and its surface area is A cm².
 	
  (a) Show that $A = 6x^2 + \dfrac{6144}{x}$.
  (b) Determine the value of x for which A has a stationary value.
  (c) By finding $\dfrac{d^2A}{dx^2}$ show that the stationary value is a minimum value.
  (d) Calculate the height of the cuboid when its surface area takes its minimum value.
	See p 9 Example 6 or review Advancing Maths for AQA C1C2 pp 233–236.

2

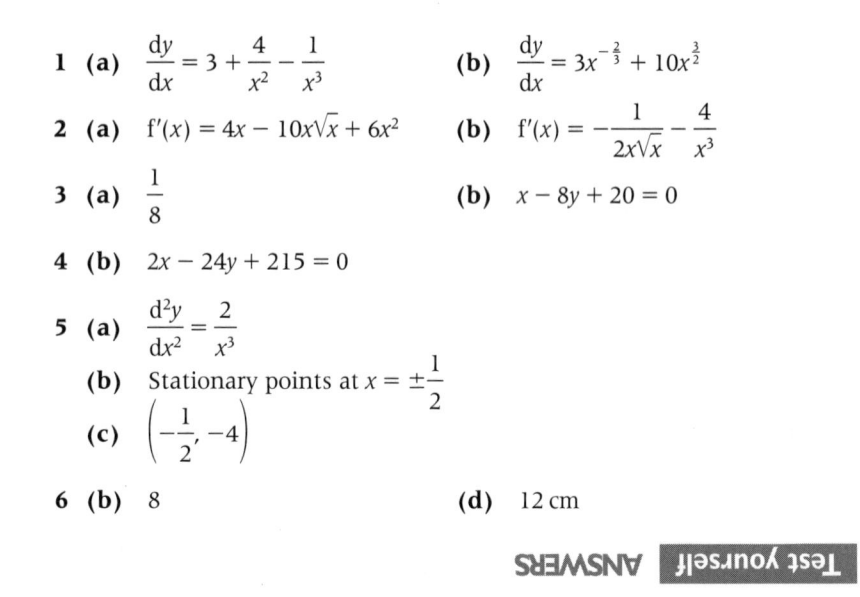

1 (a) $\dfrac{dy}{dx} = 3 + \dfrac{4}{x^2} - \dfrac{1}{x^3}$ 　　(b) $\dfrac{dy}{dx} = 3x^{-\frac{2}{3}} + 10x^{\frac{3}{2}}$

2 (a) $f'(x) = 4x - 10x\sqrt{x} + 6x^2$ 　(b) $f'(x) = -\dfrac{1}{2x\sqrt{x}} - \dfrac{4}{x^3}$

3 (a) $\dfrac{1}{8}$ 　　　　　　　　　　(b) $x - 8y + 20 = 0$

4 (b) $2x - 24y + 215 = 0$

5 (a) $\dfrac{d^2y}{dx^2} = \dfrac{2}{x^3}$

　(b) Stationary points at $x = \pm\dfrac{1}{2}$

　(c) $\left(-\dfrac{1}{2}, -4\right)$

6 (b) 8 　　　　　　　　　　　　(d) 12 cm

Further integration and the trapezium rule

3

Key points to remember

1 $\dfrac{dy}{dx} = x^n \ (n \neq -1) \implies y = \dfrac{1}{n+1}x^{n+1} + c$, where c is an arbitrary constant.

2 An indefinite integral has no limits and is of the form $\int f(x)\,dx$.

You must remember to include an arbitrary constant in your answer.

3 A definite integral is of the form $\int_b^a f(x)\,dx$, where a and b are the limits.

After integrating a definite integral, you obtain $\Big[F(x)\Big]_a^b$ and this is evaluated as $F(b) - F(a)$.

4 The area under a curve $y = f(x)$ from $x = a$ to $x = b$ is given by $\int_a^b f(x)\,dx$.

5 The trapezium rule can be used to find an approximation to a definite integral

$\int_a^b f(x)\,dx \approx \dfrac{1}{2}h\Big\{\big(y_0 + y_n\big) + 2\big(y_1 + y_2 + \dots + y_{n-1}\big)\Big\}$, where $h = \dfrac{b-a}{n}$.

Worked example 1

Given that $f'(x) = 3\sqrt{x} - \dfrac{1}{x^2}$, find an expression for $f(x)$.

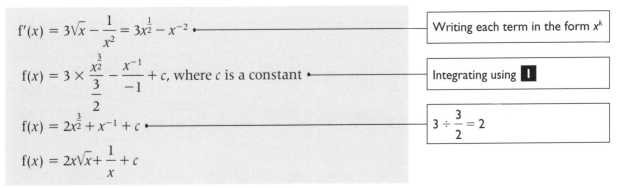

$f'(x) = 3\sqrt{x} - \dfrac{1}{x^2} = 3x^{\frac{1}{2}} - x^{-2}$ ——— Writing each term in the form x^k

$f(x) = 3 \times \dfrac{x^{\frac{3}{2}}}{\dfrac{3}{2}} - \dfrac{x^{-1}}{-1} + c$, where c is a constant ——— Integrating using **1**

$f(x) = 2x^{\frac{3}{2}} + x^{-1} + c$ ——— $3 \div \dfrac{3}{2} = 2$

$f(x) = 2x\sqrt{x} + \dfrac{1}{x} + c$

Worked example 2

A curve passes through the point $(1, 2)$ and is such that $\dfrac{dy}{dx} = \dfrac{x+4}{x^3}$.

(a) Find the equation of the curve.

(b) The curve intersects the line $x = -2$ at the point P. Find the coordinates of P.

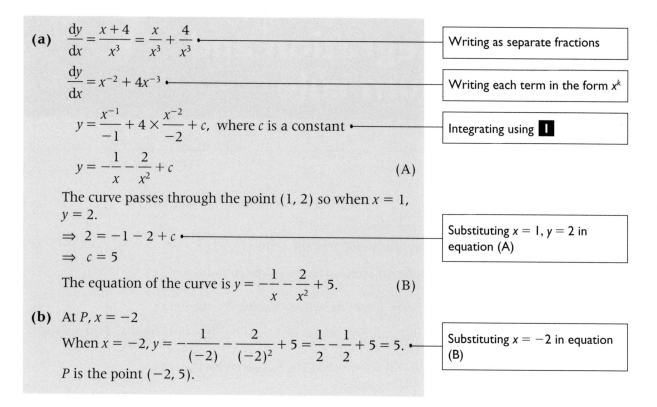

(a) $\dfrac{dy}{dx} = \dfrac{x+4}{x^3} = \dfrac{x}{x^3} + \dfrac{4}{x^3}$ — Writing as separate fractions

$\dfrac{dy}{dx} = x^{-2} + 4x^{-3}$ — Writing each term in the form x^k

$y = \dfrac{x^{-1}}{-1} + 4 \times \dfrac{x^{-2}}{-2} + c,$ where c is a constant — Integrating using ■

$y = -\dfrac{1}{x} - \dfrac{2}{x^2} + c$ \qquad (A)

The curve passes through the point $(1, 2)$ so when $x = 1$, $y = 2$.

$\Rightarrow \; 2 = -1 - 2 + c$ — Substituting $x = 1$, $y = 2$ in equation (A)

$\Rightarrow \; c = 5$

The equation of the curve is $y = -\dfrac{1}{x} - \dfrac{2}{x^2} + 5.$ \qquad (B)

(b) At P, $x = -2$

When $x = -2$, $y = -\dfrac{1}{(-2)} - \dfrac{2}{(-2)^2} + 5 = \dfrac{1}{2} - \dfrac{1}{2} + 5 = 5.$ — Substituting $x = -2$ in equation (B)

P is the point $(-2, 5)$.

Worked example 3

Find $\displaystyle\int \dfrac{(\sqrt{x} - 1)^2}{2\sqrt{x}}\, dx.$

$\displaystyle\int \dfrac{(\sqrt{x} - 1)^2}{2\sqrt{x}}\, dx = \int \dfrac{x - 2\sqrt{x} + 1}{2\sqrt{x}}\, dx$ — Using $(a - b)^2 = a^2 - 2ab + b^2$

$= \displaystyle\int \dfrac{x}{2x^{\frac{1}{2}}} - \dfrac{2x^{\frac{1}{2}}}{2x^{\frac{1}{2}}} + \dfrac{1}{2x^{\frac{1}{2}}}\, dx$ — Writing as separate fractions

$= \displaystyle\int \left(\dfrac{1}{2} \times x^{\frac{1}{2}} - 1 + \dfrac{1}{2} \times x^{-\frac{1}{2}} \right) dx$ — Writing each term in the form x^k

$= \dfrac{1}{2} \times \dfrac{1}{\frac{3}{2}} x^{\frac{3}{2}} - x + \dfrac{1}{2} \times \dfrac{1}{\frac{1}{2}} x^{\frac{1}{2}} + c,$ where c is a constant — Integrating using ■

$= \dfrac{1}{3} x^{\frac{3}{2}} - x + x^{\frac{1}{2}} + c$

$= \dfrac{1}{3} x\sqrt{x} - x + \sqrt{x} + c$

Worked example 4

Show that $\int_1^3 x^{-\frac{3}{2}}(x+1)(x-1)\,dx = \frac{8}{3}(\sqrt{3}-1)$.

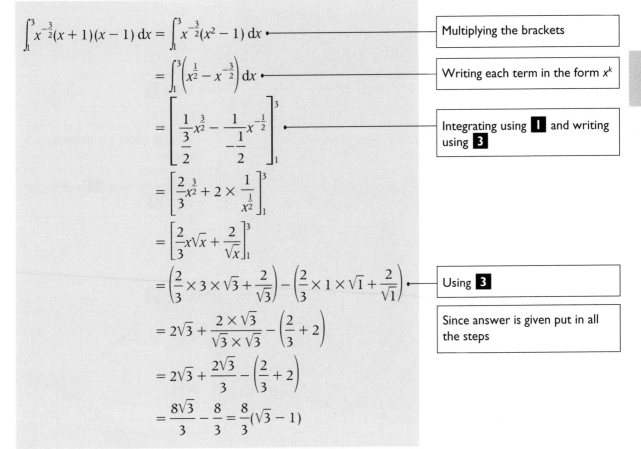

$$\int_1^3 x^{-\frac{3}{2}}(x+1)(x-1)\,dx = \int_1^3 x^{-\frac{3}{2}}(x^2-1)\,dx$$

Multiplying the brackets

$$= \int_1^3 \left(x^{\frac{1}{2}} - x^{-\frac{3}{2}}\right)dx$$

Writing each term in the form x^k

3

$$= \left[\frac{1}{\frac{3}{2}}x^{\frac{3}{2}} - \frac{1}{-\frac{1}{2}}x^{-\frac{1}{2}}\right]_1^3$$

Integrating using **1** and writing using **3**

$$= \left[\frac{2}{3}x^{\frac{3}{2}} + 2\times\frac{1}{x^{\frac{1}{2}}}\right]_1^3$$

$$= \left[\frac{2}{3}x\sqrt{x} + \frac{2}{\sqrt{x}}\right]_1^3$$

$$= \left(\frac{2}{3}\times3\times\sqrt{3} + \frac{2}{\sqrt{3}}\right) - \left(\frac{2}{3}\times1\times\sqrt{1} + \frac{2}{\sqrt{1}}\right)$$

Using **3**

Since answer is given put in all the steps

$$= 2\sqrt{3} + \frac{2\times\sqrt{3}}{\sqrt{3}\times\sqrt{3}} - \left(\frac{2}{3} + 2\right)$$

$$= 2\sqrt{3} + \frac{2\sqrt{3}}{3} - \left(\frac{2}{3} + 2\right)$$

$$= \frac{8\sqrt{3}}{3} - \frac{8}{3} = \frac{8}{3}(\sqrt{3}-1)$$

Worked example 5

The curve whose equation is $y = 9 - \dfrac{1}{x^2}$

is sketched opposite for $x > 0$. The curve intersects the positive x-axis at the point A.

(a) Find the x-coordinate of A.

(b) Calculate the area of the shaded region bounded by the curve, the x-axis and the line $x = 1$.

(c) The point $B(1, 8)$ lies on the curve. Find the area of the finite region bounded by the curve and the line AB.

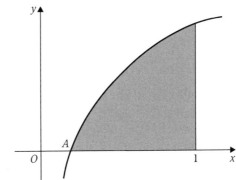

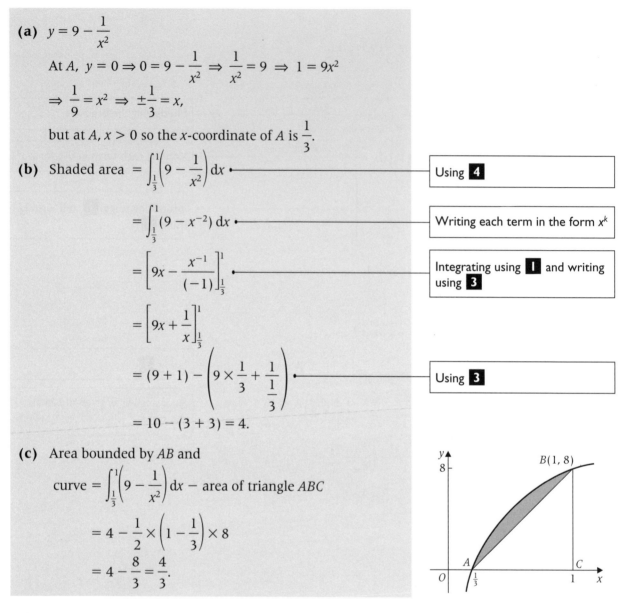

(a) $y = 9 - \dfrac{1}{x^2}$

At A, $y = 0 \Rightarrow 0 = 9 - \dfrac{1}{x^2} \Rightarrow \dfrac{1}{x^2} = 9 \Rightarrow 1 = 9x^2$

$\Rightarrow \dfrac{1}{9} = x^2 \Rightarrow \pm\dfrac{1}{3} = x,$

but at A, $x > 0$ so the x-coordinate of A is $\dfrac{1}{3}$.

(b) Shaded area $= \displaystyle\int_{\frac{1}{3}}^{1} \left(9 - \dfrac{1}{x^2}\right) dx$

> Using **4**

$\qquad = \displaystyle\int_{\frac{1}{3}}^{1} (9 - x^{-2}) \, dx$

> Writing each term in the form x^k

$\qquad = \left[9x - \dfrac{x^{-1}}{(-1)} \right]_{\frac{1}{3}}^{1}$

> Integrating using **1** and writing using **3**

$\qquad = \left[9x + \dfrac{1}{x} \right]_{\frac{1}{3}}^{1}$

$\qquad = (9 + 1) - \left(9 \times \dfrac{1}{3} + \dfrac{1}{\frac{1}{3}} \right)$

> Using **3**

$\qquad = 10 - (3 + 3) = 4.$

(c) Area bounded by AB and

curve $= \displaystyle\int_{\frac{1}{3}}^{1} \left(9 - \dfrac{1}{x^2}\right) dx -$ area of triangle ABC

$\qquad = 4 - \dfrac{1}{2} \times \left(1 - \dfrac{1}{3}\right) \times 8$

$\qquad = 4 - \dfrac{8}{3} = \dfrac{4}{3}.$

Worked example 6

(a) Use the trapezium rule with five ordinates (four strips) to find an approximate value for $\displaystyle\int_{-1}^{1} (3^{-x} + 1) \, dx$, giving your answer to three decimal places.

(b) The diagram shows a sketch of the curve with equation $y = 3^{-x} + 1$.
Explain with the aid of a diagram whether your approximate value will be an overestimate or an underestimate of the true value for $\displaystyle\int_{-1}^{1} (3^{-x} + 1) \, dx$.

(c) Comment on how you could obtain a better approximation to the value of the integral using the trapezium rule.

(a) $h = \dfrac{1-(-1)}{4} = 0.5$

Using **5** with the number of strips, *n*, being 4

x	$y = f(x) = 3^{-x} + 1$	
	ends	**middles**
$x_0 = a = -1$	$y_0 = f(-1) = 4.00000...$	
$x_1 = a + h = -0.5$		$y_1 = f(-0.5) = 2.73205...$
$x_2 = a + 2h = 0$		$y_2 = f(0) = 2.00000...$
$x_3 = a + 3h = 0.5$		$y_3 = f(0.5) = 1.57735...$
$x_4 = b = 1$	$y_4 = f(1) = 1.33333...$	
	$y_0 + y_4 = 5.33333...$	$y_1 + y_2 + y_3 = 6.30940...$ **Totals**

$$\int_{-1}^{1} (3^{-x} + 1)\, dx \approx \frac{1}{2} \times 0.5\{(5.33333...) + 2(6.30940...)\}$$

Using **5** with $n = 4$

$$\int_{-1}^{1} (3^{-x} + 1)\, dx \approx 0.25 \times \{17.95213...\} = 4.48803...$$
$$= 4.488$$
(to 3 decimal places)

(Always show working to a greater degree of accuracy and remember to give the final answer to the degree of accuracy asked for.)

(b)

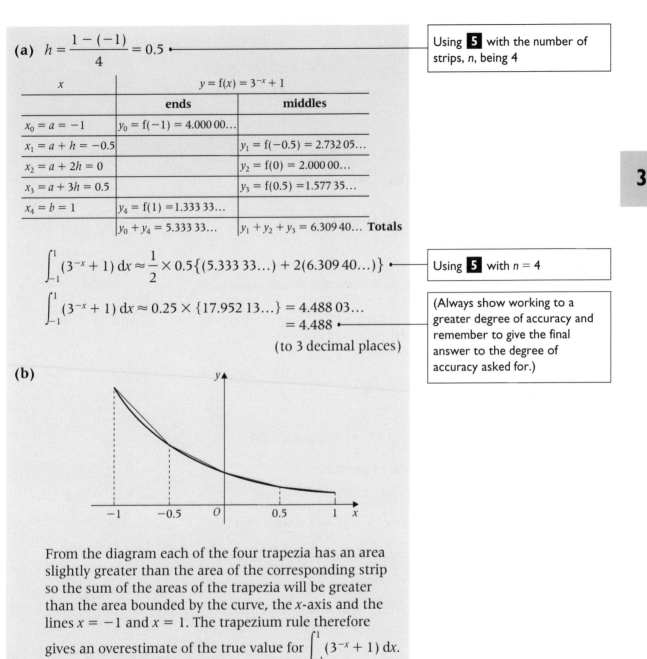

From the diagram each of the four trapezia has an area slightly greater than the area of the corresponding strip so the sum of the areas of the trapezia will be greater than the area bounded by the curve, the x-axis and the lines $x = -1$ and $x = 1$. The trapezium rule therefore gives an overestimate of the true value for $\int_{-1}^{1} (3^{-x} + 1)\, dx$.

(c) Increase the number of ordinates used in the trapezium rule.

REVISION EXERCISE 3

1 Find y in terms of x for each of the following:

(a) $\dfrac{dy}{dx} = 6x^{-7}$ **(b)** $\dfrac{dy}{dx} = 5x^{\frac{1}{4}}$ **(c)** $\dfrac{dy}{dx} = \dfrac{6}{x^4}$ **(d)** $\dfrac{dy}{dx} = \sqrt[5]{x}$

2 Find each of the following:

(a) $\displaystyle\int \left(3x^{\frac{1}{2}} + 1\right) dx$ **(b)** $\displaystyle\int \left(9x^2 - \dfrac{x^{-2}}{2}\right) dx$

(c) $\displaystyle\int x^{-4}(x^2 - 6) \, dx$ **(d)** $\displaystyle\int \dfrac{4x - 1}{2x^{\frac{1}{2}}} \, dx$

3 Find an expression for $f(x)$, in each of the following, given that:

(a) $f'(x) = \dfrac{2}{x^2} + \dfrac{3x^2}{2}$ **(b)** $f'(x) = 4\sqrt[3]{x} - \dfrac{1}{2\sqrt{x}}$

(c) $f'(x) = \dfrac{1 + x}{\sqrt{x}}$ **(d)** $f'(x) = \dfrac{(1 + x^2)(3 - 2x^2)}{x^2}$

4 A curve passes through the point $\left(3, \dfrac{2}{3}\right)$ and is such that $\dfrac{dy}{dx} = \left(x + \dfrac{1}{x}\right)^2$. Find the equation of the curve.

5 A curve passes through the point $(4, -3)$ and is such that $\dfrac{dy}{dx} = 5x(\sqrt{x} - 1)$. Find the equation of the curve.

6 A curve passes through the point $(-1, 2)$ and is such that $\dfrac{dy}{dx} = \dfrac{2 - x}{x^3} + 3$.

(a) Find the equation of the curve.

(b) The curve intersects the line $x = 1$ at the point A. Find the coordinates of A.

7 Evaluate the following integrals:

(a) $\displaystyle\int_1^4 \left(x^{-\frac{1}{2}} + 1\right) dx$ **(b)** $\displaystyle\int_2^4 \left(x + \dfrac{1}{x^2}\right) dx$

(c) $\displaystyle\int_1^2 \dfrac{5x^8 + 6}{x^4} \, dx$ **(d)** $\displaystyle\int_0^1 \sqrt[3]{x}(x - 1)^2 \, dx$

8 Find the area bounded by the curve with equation $y = 1 + \dfrac{1}{x^2}$, the x-axis and the lines $x = 1$ and $x = 2$.

9 A curve, C, is defined for $x \geqslant 0$ by the equation $y = \sqrt{x}$.

(a) Find the area of the region bounded by C, the x-axis and the lines $x = 1$ and $x = 4$.

(b) Show that the area of the finite region bounded by C, the x-axis and the line $x = 1$ is one seventh of the area of the region found in part **(a)**.

3

10 The curve whose equation is $y = 4 - \dfrac{1}{x^2}$ is sketched below for $x > 0$. The curve intersects the positive x-axis at the point A and the point $B(1, 3)$ lies on the curve.

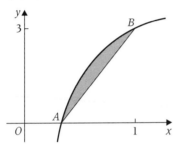

(a) Find the x-coordinate of A.

(b) Find $\displaystyle\int \left(4 - \dfrac{1}{x^2}\right) dx$.

(c) Find the area of the shaded region bounded by the curve and the line AB.

11 Use the trapezium rule with six ordinates (five strips) to find an approximate value for $\displaystyle\int_0^5 \sqrt{25 - x^2}\, dx$, giving your answer to 3 decimal places.

12 The diagram shows a sketch of the curve with equation $y = \sqrt{2x - 4}$.

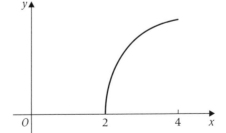

(a) Use the trapezium rule with five ordinates (four strips) to find an approximate value for $\displaystyle\int_2^4 \sqrt{2x - 4}\, dx$, giving your answer to 4 significant figures.

(b) Explain with the aid of a diagram whether your approximate value will be an overestimate or an underestimate of the true value for $\displaystyle\int_2^4 \sqrt{2x - 4}\, dx$.

13 Use the trapezium rule with five ordinates (four strips) to find an approximate value for $\int_1^3 \frac{1}{3+x^3} \, dx$, giving your answer to 4 decimal places.

14 (a) Use the trapezium rule with four ordinates (three strips) to find an approximate value for $\int_0^3 (2^x - \sqrt{x}) \, dx$, giving your answer to 3 significant figures.

(b) Comment on how you could obtain a better approximation to the value of the integral using the trapezium rule.

Test yourself	What to review
	If your answer is incorrect:
1 A curve passes through the point (4, 30) and is such that $\frac{dy}{dx} = 2x + 3\sqrt{x}.$ **(a)** Find the equation of the curve. **(b)** Verify that the point (1, 1) lies on the curve.	Sec p 15 Example 2 or review *Advancing Maths for AQA C1C2* pp 239–241.
2 Find f(x) given that f'(x) = $\frac{1 + x^6}{x^3}$.	See p 15 Example 1 or review *Advancing Maths for AQA C1C2* pp 241–242.
3 Find $\int 6\sqrt{x}(\sqrt{x} + 1)^2 \, dx$.	See p 16 Example 3 or review *Advancing Maths for AQA C1C2* pp 243–244.
4 Find the area bounded by the curve with equation $y = \frac{x + 1}{\sqrt{x}}$, the x-axis and the lines $x = 1$ and $x = 9$.	See p 17 Example 5 or review *Advancing Maths for AQA C1C2* pp 245–247.
5 (a) Use the trapezium rule with five ordinates (four strips) to find an approximate value for $\int_0^2 \sqrt{6 - x^2} \, dx$, giving your answer to 3 decimal places. **(b)** A second approximate value for $\int_0^2 \sqrt{6 - x^2} \, dx$ is found using nine ordinates (eight strips). Would you expect this value to be more accurate or less accurate than the answer to **(a)**?	See p 18 Example 6 or review *Advancing Maths for AQA C1C2* pp 247–251.

Test yourself ANSWERS

1 (a) $y = x^2 + 2x\sqrt{x} - 2$

2 $f(x) = \dfrac{x^4}{4} - \dfrac{1}{2x^2} + c$

3 $\dfrac{12}{5}x^2\sqrt{x} + 6x^2 + 4x\sqrt{x} + c$

4 $21\dfrac{1}{3}$

5 (a) 4.251 **(b)** More accurate

3

CHAPTER 4

Basic trigonometry

Key points to remember

1 The graph of $y = \sin \theta$ is periodic with period $360°$.
Vertical lines of symmetry occur at $\theta = 90°$, $\theta = 270°$, $\theta = -90°$, ..., etc.

2
$\sin (360° + \theta) = \sin \theta$ $\sin (360° - \theta) = -\sin \theta$
$\sin (180° - \theta) = \sin \theta$ $\sin (180° + \theta) = -\sin \theta$
$\sin (-\theta) = -\sin \theta$

3 From the graph of $y = \sin \theta$ you can see that two solutions of the equation $\sin \theta = \sin \alpha$, in a $360°$ cycle, are $\theta = \alpha$ and $\theta = 180° - \alpha$.
Since the graph of $y = \sin \theta$ has period $360°$, the solutions of $\sin \theta = \sin \alpha$ are

$$\theta = \alpha, \alpha + 360°, \alpha - 360°, \alpha + 720°, \alpha - 720°, ...,$$

and $\theta = 180° - \alpha$, $180° - \alpha + 360°$, $180° - \alpha - 360°$, $180° - \alpha + 720°$, $180° - \alpha - 720°$, ...

4

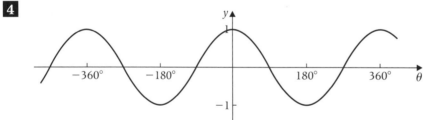

The graph of $y = \cos \theta$ is periodic with period $360°$.
Vertical lines of symmetry occur at $\theta = 0°$, $\theta = 180°$, $\theta = -180°$, ..., etc.

5
$\cos (360° + \theta) = \cos \theta$ $\cos (180° - \theta) = -\cos \theta$
$\cos (360° - \theta) = \cos \theta$ $\cos (180° + \theta) = -\cos \theta$
$\cos (-\theta) = \cos \theta$

6 From the graph of $y = \cos \theta$ you can see that two solutions of the equation $\cos \theta = \cos \alpha$, in a $360°$ cycle, are $\theta = \alpha$ and $\theta = -\alpha$.
Since the graph of $y = \cos \theta$ has period $360°$, the solutions of $\cos \theta = \cos \alpha$ are

$$\theta = \alpha, \alpha + 360°, \alpha - 360°, \alpha + 720°, \alpha - 720°, ...,$$

and $\theta = -\alpha$, $-\alpha + 360°$, $-\alpha - 360°$, $-\alpha + 720°$, $-\alpha - 720°$, ...

7

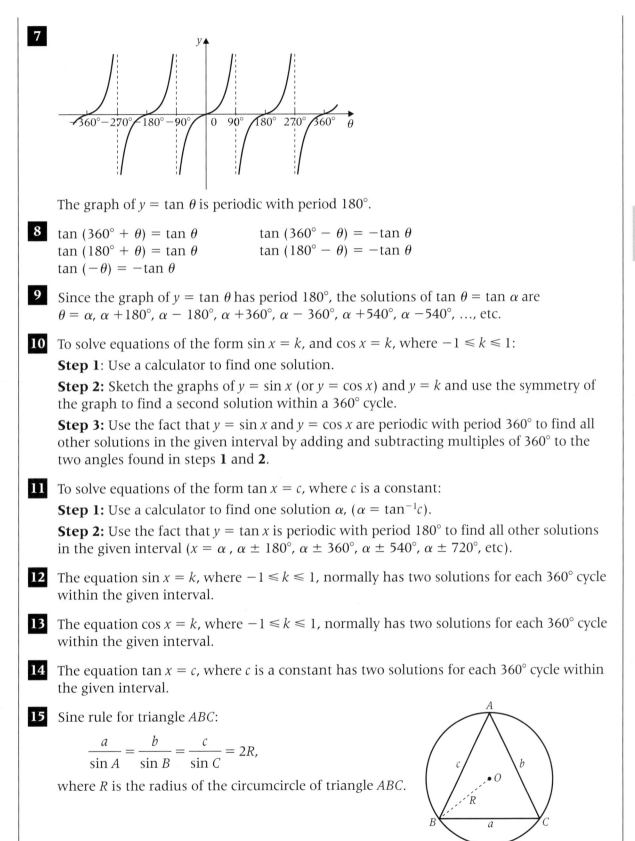

The graph of $y = \tan \theta$ is periodic with period $180°$.

8 $\tan(360° + \theta) = \tan \theta$ $\tan(360° - \theta) = -\tan \theta$
$\tan(180° + \theta) = \tan \theta$ $\tan(180° - \theta) = -\tan \theta$
$\tan(-\theta) = -\tan \theta$

9 Since the graph of $y = \tan \theta$ has period $180°$, the solutions of $\tan \theta = \tan \alpha$ are
$\theta = \alpha$, $\alpha + 180°$, $\alpha - 180°$, $\alpha + 360°$, $\alpha - 360°$, $\alpha + 540°$, $\alpha - 540°$, ..., etc.

10 To solve equations of the form $\sin x = k$, and $\cos x = k$, where $-1 \leq k \leq 1$:

Step 1: Use a calculator to find one solution.

Step 2: Sketch the graphs of $y = \sin x$ (or $y = \cos x$) and $y = k$ and use the symmetry of the graph to find a second solution within a $360°$ cycle.

Step 3: Use the fact that $y = \sin x$ and $y = \cos x$ are periodic with period $360°$ to find all other solutions in the given interval by adding and subtracting multiples of $360°$ to the two angles found in steps **1** and **2**.

11 To solve equations of the form $\tan x = c$, where c is a constant:

Step 1: Use a calculator to find one solution α, ($\alpha = \tan^{-1} c$).

Step 2: Use the fact that $y = \tan x$ is periodic with period $180°$ to find all other solutions in the given interval ($x = \alpha$, $\alpha \pm 180°$, $\alpha \pm 360°$, $\alpha \pm 540°$, $\alpha \pm 720°$, etc).

12 The equation $\sin x = k$, where $-1 \leq k \leq 1$, normally has two solutions for each $360°$ cycle within the given interval.

13 The equation $\cos x = k$, where $-1 \leq k \leq 1$, normally has two solutions for each $360°$ cycle within the given interval.

14 The equation $\tan x = c$, where c is a constant has two solutions for each $360°$ cycle within the given interval.

15 Sine rule for triangle ABC:

$$\frac{a}{\sin A} = \frac{b}{\sin B} = \frac{c}{\sin C} = 2R,$$

where R is the radius of the circumcircle of triangle ABC.

16 Cosine rule for triangle *ABC*:

$$a^2 = b^2 + c^2 - 2bc \cos A$$

To find the angles when all three sides are known use

$$\cos A = \frac{b^2 + c^2 - a^2}{2bc}.$$

17 Area of triangle $ABC = \frac{1}{2}ab \sin C$

In words:
Area of a triangle equals half the product of two of the sides times the sine of the angle between them.

Worked example 1

(a) Sketch the graph of $y = \sin x$ in the interval $0° \leqslant x \leqslant 720°$. Write down the coordinates of all the stationary points of the graph.

(b) Find all the values of x in the interval $0° \leqslant x \leqslant 720°$ for which $\sin x = \sin 72°$.

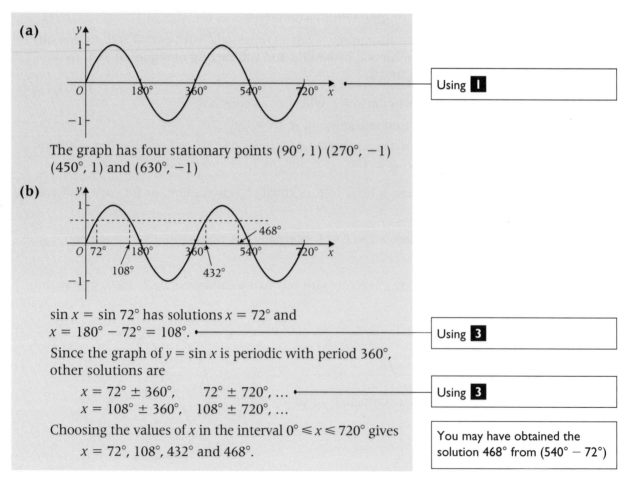

(a)

Using **1**

The graph has four stationary points $(90°, 1)$ $(270°, -1)$ $(450°, 1)$ and $(630°, -1)$

(b)

$\sin x = \sin 72°$ has solutions $x = 72°$ and $x = 180° - 72° = 108°$.

Using **3**

Since the graph of $y = \sin x$ is periodic with period $360°$, other solutions are

$$x = 72° \pm 360°, \qquad 72° \pm 720°, \dots$$
$$x = 108° \pm 360°, \quad 108° \pm 720°, \dots$$

Using **3**

Choosing the values of x in the interval $0° \leqslant x \leqslant 720°$ gives

$$x = 72°, 108°, 432° \text{ and } 468°.$$

You may have obtained the solution $468°$ from $(540° - 72°)$

Worked example 2

Solve the equation $4\cos\theta + 3 = 0$ in the interval $-360° \leqslant \theta \leqslant 360°$ giving your answers to 1 decimal place.

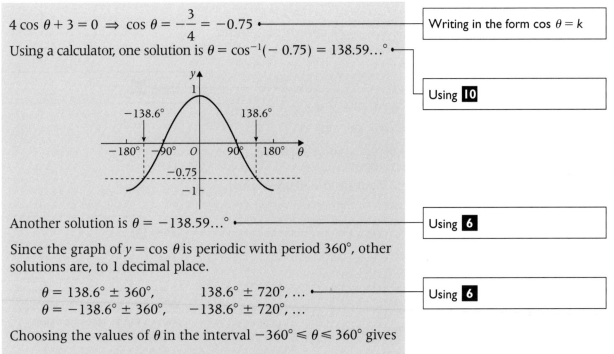

$4\cos\theta + 3 = 0 \Rightarrow \cos\theta = -\dfrac{3}{4} = -0.75$ — Writing in the form $\cos\theta = k$

Using a calculator, one solution is $\theta = \cos^{-1}(-0.75) = 138.59...°$ — Using **10**

Another solution is $\theta = -138.59...°$ — Using **6**

Since the graph of $y = \cos\theta$ is periodic with period $360°$, other solutions are, to 1 decimal place.

$\theta = 138.6° \pm 360°, \qquad 138.6° \pm 720°, ...$ — Using **6**
$\theta = -138.6° \pm 360°, \qquad -138.6° \pm 720°, ...$

Choosing the values of θ in the interval $-360° \leqslant \theta \leqslant 360°$ gives

$\theta = -221.4°, -138.6°, 138.6°$ and $221.4°$.

Worked example 3

Solve the equation $\sqrt{3}\tan\theta - 1 = 0$ in the interval $-180° < \theta < 540°$.

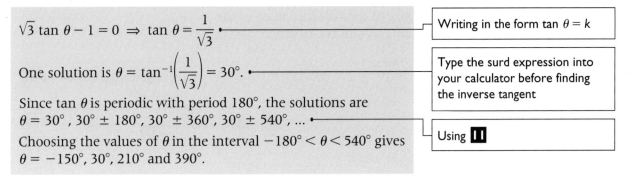

$\sqrt{3}\tan\theta - 1 = 0 \Rightarrow \tan\theta = \dfrac{1}{\sqrt{3}}$ — Writing in the form $\tan\theta = k$

One solution is $\theta = \tan^{-1}\left(\dfrac{1}{\sqrt{3}}\right) = 30°$. — Type the surd expression into your calculator before finding the inverse tangent

Since $\tan\theta$ is periodic with period $180°$, the solutions are
$\theta = 30°, 30° \pm 180°, 30° \pm 360°, 30° \pm 540°, ...$ — Using **11**

Choosing the values of θ in the interval $-180° < \theta < 540°$ gives
$\theta = -150°, 30°, 210°$ and $390°$.

Worked example 4

In triangle ABC, angle $BAC = 65°$, $AB = 8$ cm and $AC = 6$ cm.

(a) Calculate the length of BC, giving your answer to the nearest millimetre.

(b) Show that the area of triangle ABC is 22 cm² correct to the nearest cm².

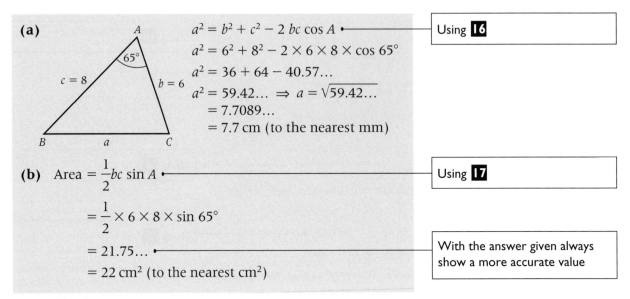

(a)

$a^2 = b^2 + c^2 - 2bc \cos A$ — Using **16**

$a^2 = 6^2 + 8^2 - 2 \times 6 \times 8 \times \cos 65°$

$a^2 = 36 + 64 - 40.57...$

$a^2 = 59.42... \Rightarrow a = \sqrt{59.42...}$

$= 7.7089...$

$= 7.7$ cm (to the nearest mm)

(b) Area $= \dfrac{1}{2}bc \sin A$ — Using **17**

$= \dfrac{1}{2} \times 6 \times 8 \times \sin 65°$

$= 21.75...$ — With the answer given always show a more accurate value

$= 22$ cm² (to the nearest cm²)

Worked example 5

The lengths of the sides of a triangle are 27 cm, 30 cm and 36 cm. The largest angle of the triangle is $\theta°$.

(a) Show that $\theta = 78.1$ correct to 3 significant figures.

(b) Use the sine rule to find the size of the smallest angle of the triangle, giving your answer to 3 significant figures.

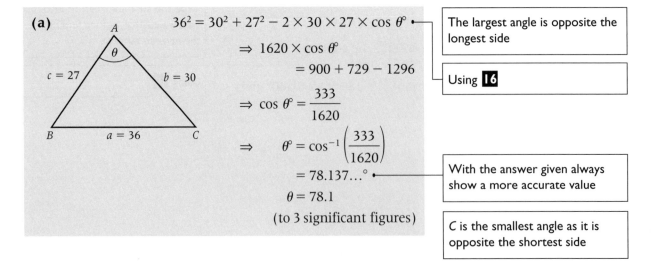

(a)

$36^2 = 30^2 + 27^2 - 2 \times 30 \times 27 \times \cos \theta°$ — The largest angle is opposite the longest side

$\Rightarrow 1620 \times \cos \theta°$

$= 900 + 729 - 1296$ — Using **16**

$\Rightarrow \cos \theta° = \dfrac{333}{1620}$

$\Rightarrow \theta° = \cos^{-1}\left(\dfrac{333}{1620}\right)$

$= 78.137...°$ — With the answer given always show a more accurate value

$\theta = 78.1$

(to 3 significant figures)

C is the smallest angle as it is opposite the shortest side

(b)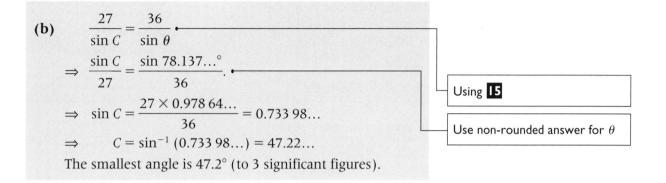

$$\frac{27}{\sin C} = \frac{36}{\sin \theta} \bullet$$

$$\Rightarrow \quad \frac{\sin C}{27} = \frac{\sin 78.137...°}{36} \bullet$$

$$\Rightarrow \quad \sin C = \frac{27 \times 0.978\,64...}{36} = 0.733\,98...$$

$$\Rightarrow \quad C = \sin^{-1}(0.733\,98...) = 47.22...$$

The smallest angle is 47.2° (to 3 significant figures).

Using **15**

Use non-rounded answer for θ

REVISION EXERCISE 4

4

1 **(a)** Sketch the graph of $y = \cos x$ in the interval
$-90° \leqslant x \leqslant 630°$. Write down the coordinates of all the
stationary points of the graph.

(b) Find all the values of x in the interval $-90° \leqslant x \leqslant 630°$
for which $\cos x = \cos 40°$.

2 Find all the values of x in the interval $-180° \leqslant x \leqslant 540°$ for
which $\sin x = \sin(-40°)$.

3 Find all the values of θ in the interval $-720° < \theta < 180°$ for
which $\tan \theta = \tan 50°$.

4 Solve the equation $\sin x = 0.7$ in the interval $0° < x < 720°$,
giving your answers to 1 decimal place.

5 Solve the equation $\tan x = -1.5$ in the interval
$-360° < x < 360°$, giving your answers to 1 decimal place.

6 Solve the equation $\cos x = -0.8$ in the interval
$-450° < x < 450°$, giving your answers to 1 decimal place.

7 Solve the equation $2 \tan \theta + 4.2 = 0$ in the interval
$-360° < \theta < 360°$, giving your answers to 1 decimal place.

8 Solve the equation $2 \sin x + 1 = 0$ in the interval
$-720° < x < 720°$.

9 Solve the equation $2 \cos x - \sqrt{3} = 0$ in the interval
$-90° < x < 900°$.

10 In triangle ABC, angle $BAC = 130°$, $AC = 6$ cm and $BC = 11$ cm.

(a) Calculate the size of the angle ABC, giving your answer to
3 significant figures.

(b) Calculate the area of triangle ABC, giving your answer to
3 significant figures

11 **(a)** Calculate the length of PR, giving your answer to the
nearest millimetre.

(b) Calculate the size of the smallest angle of the triangle,
giving your answer to 3 significant figures.

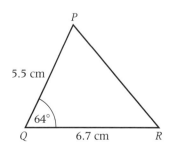

12 The lengths of the sides of a triangle are 8 cm, 9 cm and 10 cm. Calculate the size of the smallest angle of the triangle, giving your answer to 3 significant figures.

13 The area of triangle *ABC* is 24 cm².

   (a) Calculate the size of the acute angle *BAC* to the nearest 0.1°.

   (b) Calculate the length of *BC* giving your answer to the nearest millimetre.

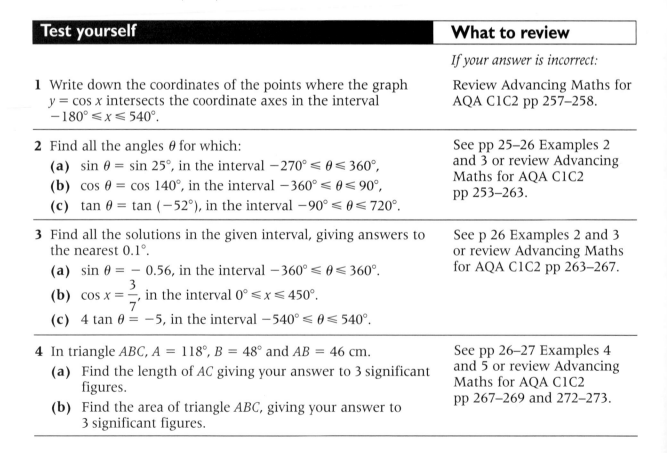

14 A circle passes through the vertices of a triangle *ABC*. Angle *ACB* = 30° and *AB* = 10 cm. The area of the circle is $k\pi$ cm², where *k* is an integer. Find the value of *k*.

15 In triangle *ABC*, angle *BAC* = 120°, *AB* = 5 cm and *AC* = 8 cm. In triangle *PQR*, angle *QPR* = 105° and *PR* = 8 cm. The length of the longest side of triangle *ABC* is the same as the length of the longest side of triangle *PQR*. Calculate the size of the smallest angle of triangle *PQR*, giving your answer to the nearest 0.1°.

16 The lengths, in centimetres, of the sides of a triangle are $(x - 1)$, x and $(x + 1)$. The largest angle of the triangle is θ.

   Show that $\cos \theta = \dfrac{x - 4}{2(x - 1)}$.

Test yourself	**What to review**
	If your answer is incorrect:
1 Write down the coordinates of the points where the graph $y = \cos x$ intersects the coordinate axes in the interval $-180° \leqslant x \leqslant 540°$.	Review Advancing Maths for AQA C1C2 pp 257–258.
2 Find all the angles θ for which: (a) $\sin \theta = \sin 25°$, in the interval $-270° \leqslant \theta \leqslant 360°$, (b) $\cos \theta = \cos 140°$, in the interval $-360° \leqslant \theta \leqslant 90°$, (c) $\tan \theta = \tan (-52°)$, in the interval $-90° \leqslant \theta \leqslant 720°$.	See pp 25–26 Examples 2 and 3 or review Advancing Maths for AQA C1C2 pp 253–263.
3 Find all the solutions in the given interval, giving answers to the nearest 0.1°. (a) $\sin \theta = -0.56$, in the interval $-360° \leqslant \theta \leqslant 360°$. (b) $\cos x = \dfrac{3}{7}$, in the interval $0° \leqslant x \leqslant 450°$. (c) $4 \tan \theta = -5$, in the interval $-540° \leqslant \theta \leqslant 540°$.	See p 26 Examples 2 and 3 or review Advancing Maths for AQA C1C2 pp 263–267.
4 In triangle *ABC*, $A = 118°$, $B = 48°$ and $AB = 46$ cm. (a) Find the length of *AC* giving your answer to 3 significant figures. (b) Find the area of triangle *ABC*, giving your answer to 3 significant figures.	See pp 26–27 Examples 4 and 5 or review Advancing Maths for AQA C1C2 pp 267–269 and 272–273.

5 In triangle *PQR*, *P* = 104°, *PQ* = 43 cm and *PR* = 52 cm. Find the perimeter of triangle *PQR* giving your answer to the nearest centimetre.

See p 26 Example 4 or review Advancing Maths for AQA C1C2 pp 271–272.

Test yourself **ANSWERS**

1 (−90°, 0) (0°, 1) (90°, 0) (270°, 0) and (450°, 0)

2 (a) −205°, 25°, 155°;

(b) −220°, −140°

(c) −52°, 128°, 308°, 488°, 668°.

3 (a) −145.9°, −34.1°, 214.1°, 325.9°

(b) 64.6°, 295.4°, 424.6°

(c) −411.3°, −231.3°, −51.3°, 128.7°, 308.7°, 488.7°

4 (a) 141 cm

(b) 2870 cm².

5 170 cm

4

CHAPTER 5

Simple transformations of graphs

Key points to remember

1 A translation of $\begin{bmatrix} 0 \\ b \end{bmatrix}$ transforms the graph of $y = f(x)$ into the graph of $y = f(x) + b$.

2 A translation of $\begin{bmatrix} a \\ 0 \end{bmatrix}$ transforms the graph of $y = f(x)$ into the graph of $y = f(x - a)$.

3 A translation of $\begin{bmatrix} a \\ b \end{bmatrix}$ transforms the graph of $y = f(x)$ into the graph of $y = f(x - a) + b$.

4 The graph of $y = f(x)$ is transformed into the graph of $y = -f(x)$ by a reflection in the line $y = 0$ (the x-axis).

5 The graph of $y = f(x)$ is transformed into the graph of $y = f(-x)$ by a reflection in the line $x = 0$ (the y-axis).

6 The graph of $y = f(x)$ is transformed into the graph of $y = d\,f(x)$ by a stretch of scale factor d in the y-direction.

7 The graph of $y = f(x)$ is transformed into the graph of $y = f\left(\dfrac{x}{c}\right)$ by a stretch of scale factor c in the x-direction.

Worked example 1

Describe the geometrical transformation that maps the graph of $y = \tan x$ onto the graph of $y = \tan(x + 60°)$

The graph of $y = \tan x$ is mapped onto the graph of $y = \tan(x + 60°)$ by a translation of $\begin{bmatrix} -60° \\ 0 \end{bmatrix}$.

Using **2** with $f(x) = \tan x$ and $a = -60°$

Worked example 2

The graph of $y = \sqrt{x}$ is translated by $\begin{bmatrix} 2 \\ -3 \end{bmatrix}$. Find the equation of the new graph.

Replace the variable x by $x - 2$ and replace the variable y with $y - (-3)$. The new graph has equation
$y + 3 = \sqrt{x - 2}$, or $y = \sqrt{x - 2} - 3$.

Using **3** with $f(x) = \sqrt{x}$ and $a = 2$ and $b = -3$

Worked example 3

The graph of $y = f(x)$ for $0 \leqslant x \leqslant 4$ is sketched opposite and $f(x) = 0$ outside of this interval.

On separate sets of axes, sketch, for $-2 \leqslant x \leqslant 6$, the graphs of:

(a) $y = f(x) + 1$ **(b)** $y = -f(x)$ **(c)** $y = f(x - 2) + 1$

(d) $y = f(-x)$ **(e)** $y = 2f(x)$

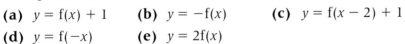

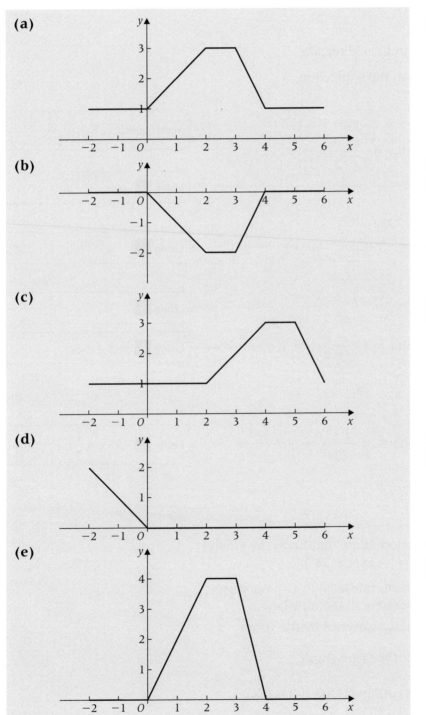

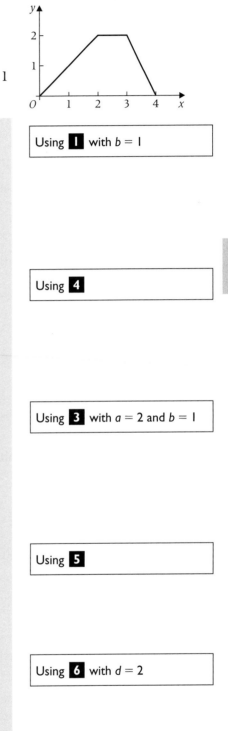

5

Worked example 4

Find the equation of the graph obtained after the graph of

$y = \dfrac{1}{2 + x}$ has been transformed by:

(a) a translation of $\begin{bmatrix} -1 \\ 0 \end{bmatrix}$,

(b) a reflection in the x-axis,

(c) a reflection in the y-axis,

(d) a stretch with scale factor 3 in the y-direction,

(e) a stretch with scale factor $\dfrac{1}{2}$ in the x-direction.

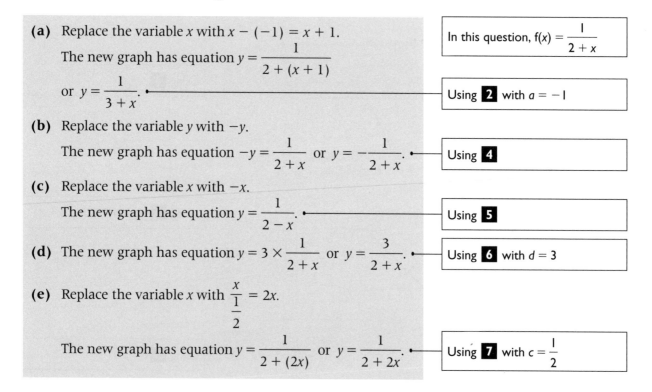

(a) Replace the variable x with $x - (-1) = x + 1$.

The new graph has equation $y = \dfrac{1}{2 + (x + 1)}$

or $y = \dfrac{1}{3 + x}$.

In this question, $f(x) = \dfrac{1}{2 + x}$

Using **2** with $a = -1$

(b) Replace the variable y with $-y$.

The new graph has equation $-y = \dfrac{1}{2 + x}$ or $y = -\dfrac{1}{2 + x}$.

Using **4**

(c) Replace the variable x with $-x$.

The new graph has equation $y = \dfrac{1}{2 - x}$.

Using **5**

(d) The new graph has equation $y = 3 \times \dfrac{1}{2 + x}$ or $y = \dfrac{3}{2 + x}$.

Using **6** with $d = 3$

(e) Replace the variable x with $\dfrac{x}{\frac{1}{2}} = 2x$.

The new graph has equation $y = \dfrac{1}{2 + (2x)}$ or $y = \dfrac{1}{2 + 2x}$.

Using **7** with $c = \dfrac{1}{2}$

REVISION EXERCISE 5

1 Describe the geometrical transformation that maps the graph of $y = \cos x$ onto the graph of $y = \cos(x + 28°)$

2 (a) The graph of $y = x^2$ has been translated to give the graph of $y = (x + 2)^2$. State the vector of the translation.

(b) The graph of $y = (x + 2)^2$ is translated by the vector $\begin{bmatrix} 0 \\ -4 \end{bmatrix}$. Find the equation of the new graph.

3 Find the equation of the graph of $y = 7^x$ after it has been reflected in the y-axis.

4 Describe the geometrical transformation that maps the graph of $y = (x + 2)^2$ onto the graph of:

   (a) $y = -(x + 2)^2$ **(b)** $y = (3x + 2)^2$

   (c) $y = (x + 1)^2 + 4$.

5 State the geometrical transformation that has taken place in mapping the graph of $y = 4 + \tan x$ onto the graph of each of the following:

   (a) $y = \tan x$ **(b)** $y = -4 - \tan x$

   (c) $y = 4 + \tan \dfrac{x}{3}$ **(d)** $y = 8 + 2 \tan x$

   (e) $y = 4 + \tan (x - 40°)$.

6 The graph of $y = x^2$ can be mapped onto the graph of $y = 16x^2$ by means of a one-way stretch. State the scale factor of the stretch when it is parallel to:

   (a) the x-axis,   **(b)** the y-axis.

7 Describe the geometrical transformation that maps the graph of $y = 1 - x^5$ onto the graph of $y = 1 + x^5$

8 Find the equation of the line $y = 2 - x$ after it has been reflected in the x-axis.

9 Find the equation of the line $y = 2 - 3x$ after it has been translated by the vector:

   (a) $\begin{bmatrix} 0 \\ -2 \end{bmatrix}$ **(b)** $\begin{bmatrix} 1 \\ 0 \end{bmatrix}$

10 Find the equation of the line $y = 2x + 4$ after it has been:

   (a) stretched in the y-direction with scale factor 2,

   (b) stretched in the x-direction with scale factor 2.

11 Describe the geometrical transformation that maps the graph of $y = \sin (x + 60°)$ onto the graph of:

   (a) $y = \sin (x + 90°)$ **(b)** $y = \sin (2x + 60°)$

   (c) $y = \dfrac{1}{5} \sin (x + 60°)$ **(d)** $y = \sin (60° - x)$.

12 The graph of $y = 4^x$ can be mapped onto the graph of $y = 4^{x + \frac{1}{2}}$ by means of either a translation or a stretch.

   (a) State the vector of the translation.

   (b) Describe the stretch in detail.

13 The graph of $y = f(x)$ for $0 \leqslant x \leqslant 2$ is sketched opposite and $f(x) = 0$ outside of this interval.

On separate sets of axes, sketch, for $-2 \leqslant x \leqslant 3$, the graphs of:

   (a) $y = -f(x)$ **(b)** $y = f(-x)$ **(c)** $y = f(x - 1)$

   (d) $y = 2f(x)$ **(e)** $y = f(2x)$ **(f)** $y = f\left(\dfrac{2}{3}x\right)$.

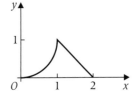

Test yourself	What to review
	If your answer is incorrect:
1 Describe geometrically the transformation that maps the graph of $y = x^5$ onto the graph of: **(a)** $y = x^5 - 4$ **(b)** $y = (x + 7)^5$ **(c)** $y = (x - 1)^5 + 3$.	See p 30 Examples 1 and 2 or review Advancing Maths for AQA C1C2 pp 279–282.
2 Find the equation of the resulting graph when the graph of $y = 3 - x^3$ is reflected: **(a)** in the *x*-axis, **(b)** in the *y*-axis.	See p 32 Example 4 or review Advancing Maths for AQA C1C2 pp 282–283.
3 State the geometrical transformation that maps the graph of $y = \cos x$ onto the graph of $y = 8 \cos x$.	See p 31 Example 3 or review Advancing Maths for AQA C1C2 p 283.
4 Find the equation of the resulting graph when the graph of $y = \cos x$ is stretched by a scale factor 6 in the *x*-direction.	See p 32 Example 4 or review Advancing Maths for AQA C1C2 p284.

Test yourself ANSWERS

4 $y = \cos \dfrac{x}{6}$

3 A stretch in the *y*-direction with scale factor 8

2 (a) $y = x^3 - 3$ **(b)** $y = 3 + x^3$

1 (a) A translation of $\begin{bmatrix} 0 \\ -4 \end{bmatrix}$ **(b)** A translation of $\begin{bmatrix} -7 \\ 0 \end{bmatrix}$

(c) A translation of $\begin{bmatrix} 1 \\ 3 \end{bmatrix}$

CHAPTER 6

Solving trigonometrical equations

Key points to remember

1 To solve equations of the form $\sin bx = k$, $\cos bx = k$ and $\tan bx = c$ in a given interval for x:

Step 1: Let $u = bx$, to get an equation of the form $\sin u = k$, $\cos u = k$ and $\tan u = c$

Step 2: Find the interval for u.

Step 3: Solve $\sin u = k$, $\cos u = k$ and $\tan u = c$ using the symmetries of their graphs as in chapter 4 to get the values of u in the interval for u.

Step 4: Substitute bx for u and hence obtain the values of x.

2 To solve equations of the form $\sin (x + b) = k$, $\cos (x + b) = k$ and $\tan (x + b) = c$ in a given interval for x:

Step 1: Let $u = (x + b)$, to get an equation of the form $\sin u = k$, $\cos u = k$ and $\tan u = c$.

Step 2: Find the interval for u.

Step 3: Solve $\sin u = k$, $\cos u = k$ and $\tan u = c$ using the symmetries of their graphs as in chapter 4 to get the values of u in the interval for u.

Step 4: Substitute $x + b$ for u and hence obtain the values of x.

3 To solve a quadratic equation in $\sin x$ in a given interval for x:

Step 1: Factorise or use the quadratic formula to find the value(s) (normally two different values) of $\sin x$, for example $\sin x = p$ or $\sin x = q$.

Step 2: Consider each of the equations $\sin x = p$ and $\sin x = q$ separately to find all the solutions for x in the given interval.

[The same method is used to solve quadratic equations in $\cos x$ and in $\tan x$.]

4 The identity $\cos^2\theta + \sin^2\theta \equiv 1$ is frequently used to simplify equations into quadratics in either $\sin \theta$ or in $\cos \theta$.

5 The identity $\tan \theta \equiv \dfrac{\sin \theta}{\cos \theta}$, where $\cos \theta \neq 0$, is frequently used to solve equations of the form $p \sin b\theta = q \cos b\theta$, where p, q and b are constants.

6 To prove an identity, consider the expression on one side and try to reduce it to the expression on the other side. If this proves to be too difficult then try to show that the RHS and the LHS are equivalent to a third expression.

Worked example 1

Solve the equation $\sin \dfrac{x}{2} = 0.5$ in the interval $0° \leqslant x \leqslant 1440°$.

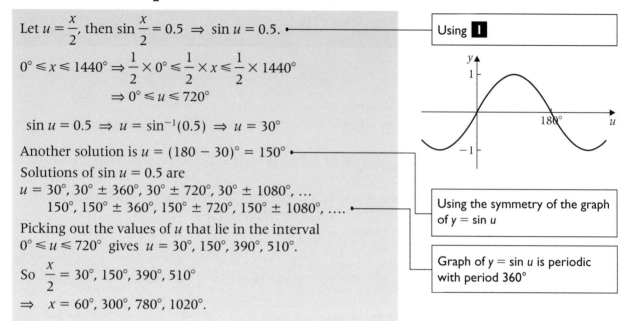

Let $u = \dfrac{x}{2}$, then $\sin \dfrac{x}{2} = 0.5 \Rightarrow \sin u = 0.5$.

Using **1**

$0° \leqslant x \leqslant 1440° \Rightarrow \dfrac{1}{2} \times 0° \leqslant \dfrac{1}{2} \times x \leqslant \dfrac{1}{2} \times 1440°$

$\Rightarrow 0° \leqslant u \leqslant 720°$

$\sin u = 0.5 \Rightarrow u = \sin^{-1}(0.5) \Rightarrow u = 30°$

Another solution is $u = (180 - 30)° = 150°$

Using the symmetry of the graph of $y = \sin u$

Solutions of $\sin u = 0.5$ are
$u = 30°, 30° \pm 360°, 30° \pm 720°, 30° \pm 1080°, \dots$
$\quad 150°, 150° \pm 360°, 150° \pm 720°, 150° \pm 1080°, \dots$.

Graph of $y = \sin u$ is periodic with period $360°$

Picking out the values of u that lie in the interval
$0° \leqslant u \leqslant 720°$ gives $u = 30°, 150°, 390°, 510°$.

So $\dfrac{x}{2} = 30°, 150°, 390°, 510°$

$\Rightarrow x = 60°, 300°, 780°, 1020°$.

Worked example 2

Solve the equation $4 \tan(\theta + 42°) + 1 = 0$ in the interval
$-360° \leqslant \theta \leqslant 360°$ giving your answers to the nearest degree.

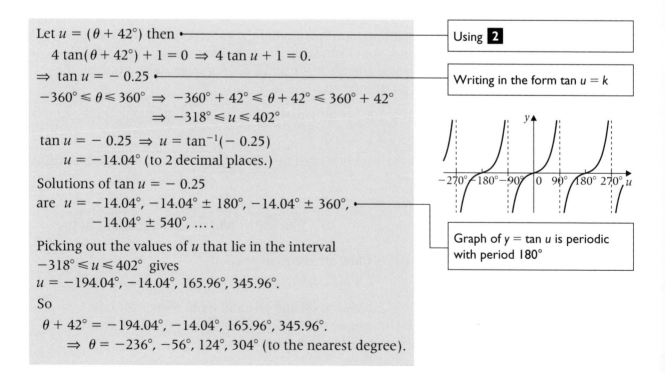

Let $u = (\theta + 42°)$ then

Using **2**

$4 \tan(\theta + 42°) + 1 = 0 \Rightarrow 4 \tan u + 1 = 0$.

$\Rightarrow \tan u = -0.25$

Writing in the form $\tan u = k$

$-360° \leqslant \theta \leqslant 360° \Rightarrow -360° + 42° \leqslant \theta + 42° \leqslant 360° + 42°$
$\quad\quad \Rightarrow -318° \leqslant u \leqslant 402°$

$\tan u = -0.25 \Rightarrow u = \tan^{-1}(-0.25)$
$\quad u = -14.04°$ (to 2 decimal places.)

Solutions of $\tan u = -0.25$
are $u = -14.04°, -14.04° \pm 180°, -14.04° \pm 360°,$
$\quad\quad -14.04° \pm 540°, \dots$.

Graph of $y = \tan u$ is periodic with period $180°$

Picking out the values of u that lie in the interval
$-318° \leqslant u \leqslant 402°$ gives
$u = -194.04°, -14.04°, 165.96°, 345.96°$.
So
$\quad \theta + 42° = -194.04°, -14.04°, 165.96°, 345.96°$.
$\quad\quad \Rightarrow \theta = -236°, -56°, 124°, 304°$ (to the nearest degree).

Worked example 3

Solve the equation $2\sin^2 x + \cos x - 1 = 0$ in the interval $-360° < x \leqslant 720°$.

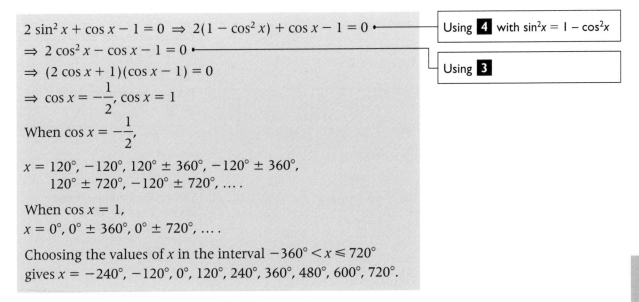

$$2\sin^2 x + \cos x - 1 = 0 \Rightarrow 2(1 - \cos^2 x) + \cos x - 1 = 0$$

Using **4** with $\sin^2 x = 1 - \cos^2 x$

$$\Rightarrow 2\cos^2 x - \cos x - 1 = 0$$

Using **3**

$$\Rightarrow (2\cos x + 1)(\cos x - 1) = 0$$

$$\Rightarrow \cos x = -\frac{1}{2}, \cos x = 1$$

When $\cos x = -\frac{1}{2}$,

$$x = 120°, -120°, 120° \pm 360°, -120° \pm 360°,$$
$$120° \pm 720°, -120° \pm 720°, \dots .$$

When $\cos x = 1$,
$$x = 0°, 0° \pm 360°, 0° \pm 720°, \dots .$$

Choosing the values of x in the interval $-360° < x \leqslant 720°$ gives $x = -240°, -120°, 0°, 120°, 240°, 360°, 480°, 600°, 720°$.

6

Worked example 4

Prove the identity $\left(\dfrac{1}{\cos x} - \tan x\right)\left(\dfrac{1}{\cos x} + \tan x\right) = 1$.

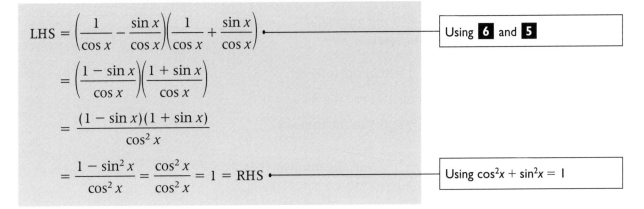

$$\text{LHS} = \left(\frac{1}{\cos x} - \frac{\sin x}{\cos x}\right)\left(\frac{1}{\cos x} + \frac{\sin x}{\cos x}\right)$$

Using **6** and **5**

$$= \left(\frac{1 - \sin x}{\cos x}\right)\left(\frac{1 + \sin x}{\cos x}\right)$$

$$= \frac{(1 - \sin x)(1 + \sin x)}{\cos^2 x}$$

$$= \frac{1 - \sin^2 x}{\cos^2 x} = \frac{\cos^2 x}{\cos^2 x} = 1 = \text{RHS}$$

Using $\cos^2 x + \sin^2 x = 1$

Worked example 5

Solve the equation $\sin 3\theta (\sin \theta - 2 \cos \theta) = 0$ in the interval $0° < \theta \leqslant 360°$, giving your answers, where appropriate, to the nearest 0.1°.

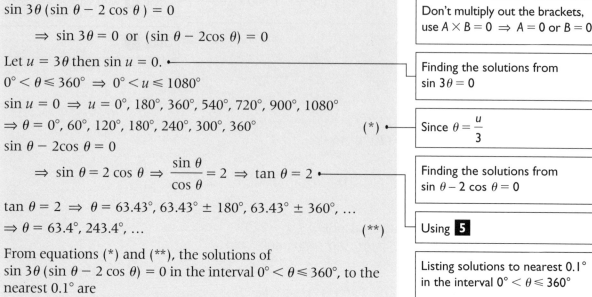

$\sin 3\theta (\sin \theta - 2 \cos \theta) = 0$

$\Rightarrow \sin 3\theta = 0$ or $(\sin \theta - 2\cos \theta) = 0$

Don't multiply out the brackets, use $A \times B = 0 \Rightarrow A = 0$ or $B = 0$

Let $u = 3\theta$ then $\sin u = 0$. •————

$0° < \theta \leqslant 360° \Rightarrow 0° < u \leqslant 1080°$

$\sin u = 0 \Rightarrow u = 0°, 180°, 360°, 540°, 720°, 900°, 1080°$

$\Rightarrow \theta = 0°, 60°, 120°, 180°, 240°, 300°, 360°$ (*) •

$\sin \theta - 2\cos \theta = 0$

$\Rightarrow \sin \theta = 2 \cos \theta \Rightarrow \dfrac{\sin \theta}{\cos \theta} = 2 \Rightarrow \tan \theta = 2$ •

$\tan \theta = 2 \Rightarrow \theta = 63.43°, 63.43° \pm 180°, 63.43° \pm 360°, \ldots$

$\Rightarrow \theta = 63.4°, 243.4°, \ldots$ (**)

Finding the solutions from $\sin 3\theta = 0$

Since $\theta = \dfrac{u}{3}$

Finding the solutions from $\sin \theta - 2 \cos \theta = 0$

Using **5**

From equations (*) and (**), the solutions of $\sin 3\theta (\sin \theta - 2 \cos \theta) = 0$ in the interval $0° < \theta \leqslant 360°$, to the nearest 0.1° are

Listing solutions to nearest 0.1° in the interval $0° < \theta \leqslant 360°$

$\theta = 0°, 60°, 63.4°, 120°, 180°, 240°, 243.4°, 300°$ and $360°$.

REVISION EXERCISE 6

1 Solve the equation $\tan 3x = 1$ in the interval $0° \leqslant x < 360°$.

2 Solve the equation $\sin 2x = 0.7$ in the interval $0° < x < 360°$, giving your answers to 1 decimal places.

3 Solve the equation $4 \cos \dfrac{1}{3}x + 1 = 0$ in the interval $-540° < x < 540°$, giving your answers to the nearest 0.1°.

4 Solve the equation $\cos(x + 32°) = 0.5$ in the interval $-360° < x < 360°$.

5 Solve the equation $5 \tan(\theta - 65°) = 9$ in the interval $-300° < \theta < 270°$, giving your answers to 1 decimal place.

6 Solve the equation $10 \sin(\theta + 20°) + 9 = 0$ in the interval $-65° \leqslant \theta \leqslant 360°$, giving your answers to 1 decimal place.

7 Solve the equation $\sin x(\sin x + 1) = 0$ in the interval $-360° \leqslant x \leqslant 720°$.

8 Solve the equation $\cos^2 x - \cos x = 0$ in the interval $-90° \leqslant x \leqslant 360°$.

9 Solve the equation $\tan^2 x + \tan x = 0$ in the interval $-180° \leqslant x \leqslant 360°$.

10 It is given that $\sin x = \dfrac{5}{13}$.

    (a) Show that the two possible values of $\cos x$ are $\pm\dfrac{12}{13}$.

    (b) Write down, in an exact form, the two possible values of $\tan x$.

11 It is given that $2 \sin 3x - 4 \cos 3x = 0$.

    (a) Write down the value of $\tan 3x$.

    (b) Hence solve the equation $2 \sin 3x - 4 \cos 3x = 0$ in the interval $-100° \leqslant x \leqslant 180°$, giving your answers to the nearest degree.

12 It is given that $2 \cos^2 \theta + 7 \sin \theta + 2 = 0$.

    (a) Show that $2x^2 - 7x - 4 = 0$, where $x = \sin \theta$.

    (b) **(i)** Solve $2x^2 - 7x - 4 = 0$.

        (ii) Hence find all solutions of the equation $2 \cos^2 \theta + 7 \sin \theta + 2 = 0$, in the interval $-360° \leqslant \theta \leqslant 360°$.

13 Prove the identity $\cos^4 x - \sin^4 x = \cos^2 x - \sin^2 x$.

14 **(a)** Find, in surd form, the solutions of the equation $x^2 - 2x - 2 = 0$.

    (b) Solve the equation $\sin^2 \theta + 2 \cos \theta + 1 = 0$ in the interval $0° \leqslant \theta \leqslant 360°$, giving your answers to the nearest $0.1°$.

15 Solve the equation $\cos x(3 \cos 2x + 5 \sin 2x) = 0$ in the interval $-180° \leqslant x \leqslant 180°$, giving your answers to the nearest degree.

16 Find all solutions of the equation $8 \tan \theta + 3 \cos \theta = 0$, in the interval $-360° \leqslant \theta \leqslant 360°$, giving your answers to the nearest $0.1°$.

6

Test yourself	**What to review**
	If your answer is incorrect:
1 Solve the following equations, giving your answers to the nearest $0.1°$.	See p 36 Example 1 or review Advancing Maths for AQA C1C2 pp 288–291.
(a) $\tan 2\theta = 0.56$, in the interval $-180° \leqslant \theta \leqslant 180°$.	
(b) $5 \cos \dfrac{x}{2} + 3 = 0$ in the interval $-360° \leqslant x \leqslant 360°$.	
2 Find all the solutions in the given interval:	See p 36 Example 2 or review Advancing Maths for AQA C1C2 pp 291–292.
(a) $\sin(\theta - 25°) = -0.5$, in the interval $-65° \leqslant \theta \leqslant 360°$,	
(b) $\tan(x - 15°) = 1$, in the interval $-360° \leqslant x \leqslant 360°$.	

Test yourself (continued)	What to review
3 Solve $2\cos^2 x + 7\sin x - 5 = 0$, in the interval $-540° \leqslant x \leqslant 540°$.	See p 37 Example 3 or review Advancing Maths for AQA C1C2 pp 292–297.
4 Show that $\dfrac{1}{\cos^2 \theta} - 1 = \tan^2 \theta$, for all values of θ.	See p 37 Example 4 or review Advancing Maths for AQA C1C2 pp 295–297.
5 Solve the equation $2\sin 2x - 3\cos 2x = 0$, giving answers in the interval $-90° \leqslant x \leqslant 180°$ to the nearest $0.1°$	See p 38 Example 5 or review Advancing Maths for AQA C1C2 pp 295–297.

Test yourself **ANSWERS**

1 (a) $-165.4°, -75.4°, 14.6°, 104.6°$ **(b)** $-253.7°, 253.7°$

2 (a) $-5°, 235°, 355°$ **(b)** $-300°, -120°, 60°, 240°$

3 $-330°, -210°, 30°, 150°, 390°, 510°$

5 $-61.8°, 28.2°, 118.2°$

CHAPTER 7

Factorials and binomial expansions

Key points to remember

1 The expression $n!$ is read as 'n factorial' and its value is given by the product $n \times (n - 1) \times (n - 2) \times \ldots \times 3 \times 2 \times 1$.

2 $n! = n \times (n - 1)!$

3 $0! = 1$

4 In general $\binom{n}{r} = \dfrac{n!}{r!(n - r)!}$ and represents the number of ways that r different objects can be chosen from n objects.

5 In general $\binom{n}{r} = \binom{n}{n - r}$.

6 The first rows of Pascal's Triangle are

$$
\begin{array}{ccccccccccccc}
& & & & & & 1 & & & & & & \\
& & & & & 1 & & 1 & & & & & \\
& & & & 1 & & 2 & & 1 & & & & \\
& & & 1 & & 3 & & 3 & & 1 & & & \\
& & 1 & & 4 & & 6 & & 4 & & 1 & & \\
& 1 & & 5 & & 10 & & 10 & & 5 & & 1 & \\
1 & & 6 & & 15 & & 20 & & 15 & & 6 & & 1
\end{array}
$$

7 Pascal's Triangle is useful for expanding $(a + b)^n$ when n is a small positive integer.

8 The general binomial expansion can be written as

$$(a + b)^n = a^n + \binom{n}{1}a^{n-1}b + \binom{n}{2}a^{n-2}b^2 + \ldots + \binom{n}{r}a^{n-r}b^r + \ldots + b^n$$

where n is a positive integer.

Worked example 1

Find the value of $\binom{8}{0} + \binom{8}{3}$.

$$\binom{8}{0} + \binom{8}{3} = \frac{8!}{0!(8 - 0)!} + \frac{8!}{3!(8 - 3)!}$$ — Using **4**

$$= \frac{8!}{1 \times 8!} + \frac{8!}{3! \times 5!}$$ — Using **3**

$$= 1 + \frac{8 \times 7 \times 6 \times 5!}{3! \times 5!}$$ — Using **2**

$$= 1 + \frac{8 \times 7 \times 6}{3 \times 2 \times 1}$$ — Using **1**

$$= 1 + 56 = 57$$

Worked example 2

(a) Use Pascal's Triangle to find the expansion of $(2 + 3x)^4$.

(b) Hence write down the expansion of $(2 - 3x)^4$.

(a) Since the index is 4, use the row of Pascal's Triangle with values

$$\begin{array}{ccccc} 1 & 4 & 6 & 4 & 1 \end{array}$$

— Using **6**

$(2 + 3x)^4 = 1 \times 2^4 \times (3x)^0 + 4 \times 2^3 \times (3x)^1 + 6 \times 2^2 \times (3x)^2$
$\qquad\qquad + 4 \times 2^1 \times (3x)^3 + 1 \times 2^0 \times (3x)^4$

— Putting $3x$ in brackets

$\qquad = 16 \times (3x)^0 + 32 \times (3x) + 24 \times (3x)^2 + 8 \times (3x)^3$
$\qquad\qquad + 1 \times (3x)^4$

— Since $2^0 = 1$

$\qquad = 16 + 96x + 24 \times (9x^2) + 8 \times (27x^3) + (81x^4)$

— Since $(3x)^0 = 1$

$\qquad = 16 + 96x + 216x^2 + 216x^3 + 81x^4$

(b) $(2 - 3x)^4 = [2 + 3(-x)]^4$
$\qquad = 16 + 96(-x) + 216(-x)^2 + 216(-x)^3 + 81(-x)^4$

— Using **(a)** with x replaced by $-x$

$\qquad = 16 - 96x + 216x^2 - 216x^3 + 81x^4$

Worked example 3

(a) Find the binomial expansion of $(2 + kx)^9$, where k is a positive constant, in ascending powers of x up to and including the term in x^2.

(b) Hence, given that the coefficient of x^2 in the expansion of $(2 + kx)^9$ is 72, find the value of k.

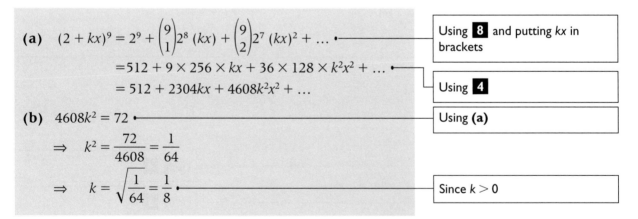

(a) $(2 + kx)^9 = 2^9 + \binom{9}{1} 2^8 (kx) + \binom{9}{2} 2^7 (kx)^2 + \dots$

— Using **8** and putting kx in brackets

$\qquad = 512 + 9 \times 256 \times kx + 36 \times 128 \times k^2 x^2 + \dots$

— Using **4**

$\qquad = 512 + 2304kx + 4608k^2 x^2 + \dots$

(b) $4608k^2 = 72$

— Using **(a)**

$\Rightarrow \quad k^2 = \dfrac{72}{4608} = \dfrac{1}{64}$

$\Rightarrow \quad k = \sqrt{\dfrac{1}{64}} = \dfrac{1}{8}$

— Since $k > 0$

Worked example 4

(a) Find the binomial expansion of $\left(1 - \frac{1}{2}x\right)^{10}$ in ascending powers of x up to and including the term in x^3.

(b) Find the coefficient of x^2 in the expansion of $(2 + x)^8$.

(c) Find the coefficient of x^2 in the expansion of
$$(2 + x)^8 \left(1 - \frac{1}{2}x\right)^{10}.$$

(a) $\left(1 - \frac{1}{2}x\right)^{10} = \left[1 + \left(-\frac{1}{2}x\right)\right]^{10}$

$= 1^{10} + \binom{10}{1} \times 1^9 \times \left(-\frac{1}{2}x\right)^1 + \binom{10}{2} \times 1^8 \times \left(-\frac{1}{2}x\right)^2$

$+ \binom{10}{3} \times 1^7 \times \left(-\frac{1}{2}x\right)^3 + \dots$ ← Using **8**

$= 1 + \frac{10!}{1! \times 9!} \times \left(-\frac{1}{2}x\right) + \frac{10!}{2! \times 8!} \times \left(\frac{1}{4}x^2\right) + \frac{10!}{3! \times 7!} \times \left(-\frac{1}{8}x^3\right) + \dots$ ← Using **4**

$= 1 + 10\left(-\frac{1}{2}x\right) + \frac{10 \times 9}{2 \times 1} \times \left(\frac{1}{4}x^2\right) + \frac{10 \times 9 \times 8}{3 \times 2 \times 1} \times \left(-\frac{1}{8}x^3\right) + \dots$ ← Using **2** and **1**

$\Rightarrow \left(1 - \frac{1}{2}x\right)^{10} = 1 - 5x + \frac{45}{4}x^2 - 15x^3 + \dots$

(b) The x^2 term in the expansion of $(2 + x)^8$ is $\binom{8}{2} \times 2^6 x^2$ ← Using **8**

$= \frac{8!}{2! \times 6!} \times 2^6 x^2$ ← Using **4**

$= \frac{8 \times 7 \times 6!}{2 \times 1 \times 6!} \times 2^6 x^2$ ← Using **2** and **1**

The coefficient of x^2 is $28 \times 2^6 = 1792$.

(c) $(2 + x)^8 \left(1 - \frac{1}{2}x\right)^{10}$

$= \left[2^8 + \binom{8}{1}2^7 x + 1792 x^2 + \dots\right]\left[1 - 5x + \frac{45}{4}x^2 - 15x^3 + \dots\right]$ ← Using **8** and the previous two parts

The x^2 terms are $2^8 \times \frac{45}{4}x^2 + \binom{8}{1}2^7 \times (-5)x^2 + 1792 x^2$

$= (2880 - 5120 + 1792)x^2$ ← Using **4** and **2**

$= -448 x^2$

The coefficient of x^2 is -448.

7

REVISION EXERCISE 7

1 Find the value of each of the following:

    (a) $6!$ **(b)** $2! + 0!$ **(c)** $\dfrac{100!}{99!}$

2 Find the value of $\dbinom{9}{5} + \dbinom{9}{4}$.

3 Use Pascal's Triangle, or otherwise, to find the expansions of:

    (a) $(1 + x)^5$ **(b)** $(2 - x)^4$ **(c)** $(a + 2b)^3$

4 (a) Obtain the expansion of $(2a + b)^4$.

    (b) Hence write down the expansion of $(b - 2a)^4$.

    (c) Hence, or otherwise, show that
        $(2a + b)^4 - (b - 2a)^4 = 16ab(4a^2 + b^2)$.

5 (a) Expand $(1 + 2x)^3$.

    (b) Write down the expansion of $(1 - 2x)^3$.

    (c) Hence solve the equation $(1 + 2x)^3 + (1 - 2x)^3 = 1178$.

6 (a) Expand $(1 + x)^4$.

    (b) Hence express $(1 + \sqrt{2})^4$ in the form $p + q\sqrt{2}$, where p and q are integers.

7 Find the coefficient of x^8 in the binomial expansion of each of the following:

    (a) $(1 + x)^{14}$ **(b)** $\left(2 - \dfrac{1}{2}x\right)^{15}$

8 The coefficient of x^{10} in the expansion of $(k + x)^{12}$ is 594. Find the possible values of the constant k.

9 Express $(\sqrt{2} - 1)^6$ in the form $a + b\sqrt{2}$.

10 (a) Find the binomial expansion of $(1 - 4x)^{10}$ in ascending powers of x up to and including the term in x^3.

    (b) Hence find the coefficient of x^3 in the expansion of
        $\left(1 + \dfrac{3}{4}x\right)(1 - 4x)^{10}$.

11 (a) Express $(1 + 3x)^5$ in the form $1 + ax + bx^2 + cx^3 + dx^4 + ex^5$.

    (b) Find the coefficient of x^8 in the expansion of $(4 - x)^9$.

    (c) Find the coefficient of x^{13} in the expansion of
        $(4 - x)^9(1 + 3x)^5$.

Test yourself	**What to review**
	If your answer is incorrect:
1 Find the value of: (a) 5! (b) $\dfrac{4!}{0!}$ (c) $\dbinom{7}{4} + \dbinom{7}{3}$.	See p 41 Example 1 or review Advancing Maths for AQA C1C2 pp 303–306.
2 Use Pascal's Triangle to find the expansion of $(2x - y)^3$.	See p 42 Example 2 or review Advancing Maths for AQA C1C2 pp 306–310.
3 Find the coefficient of p^5 in the binomial expansion of: (a) $(1 + p)^5$ (b) $(2 + p)^6$ (c) $(p - 3)^7$	See p 43 Example 4 or review Advancing Maths for AQA C1C2 pp 310–312.
4 (a) Find the expansion of $(1 + x)^8$. (b) Hence show that $\dfrac{(1 + x)^8 - (1 - x)^8}{16} = x + 7x^3 + 7x^5 + x^7$.	See p 43 Example 4 or review Advancing Maths for AQA C1C2 pp 310–315.
5 (a) Find the first four terms in ascending powers of x in the binomial expansion of $\left(2 - \dfrac{1}{4}x\right)^7$. (b) Hence find the coefficient of x^3 in the expansion of $(1 + x)^2\left(2 - \dfrac{1}{4}x\right)^7$.	See p 42 Example 3 or review Advancing Maths for AQA C1C2 pp 310–315.

7

Test yourself **ANSWERS**

5 (a) $128 - 112x + 42x^2 - 8.75x^3$ **(b)** -36.75

4 (a) $1 + 8x + 28x^2 + 56x^3 + 70x^4 + 56x^5 + 28x^6 + 8x^7 + x^8$

3 (a) 1 **(b)** 12 **(c)** 189

2 $8x^3 - 12x^2y + 6xy^2 - y^3$

1 (a) 120 **(b)** 24 **(c)** 70

CHAPTER 8

Sequences and series

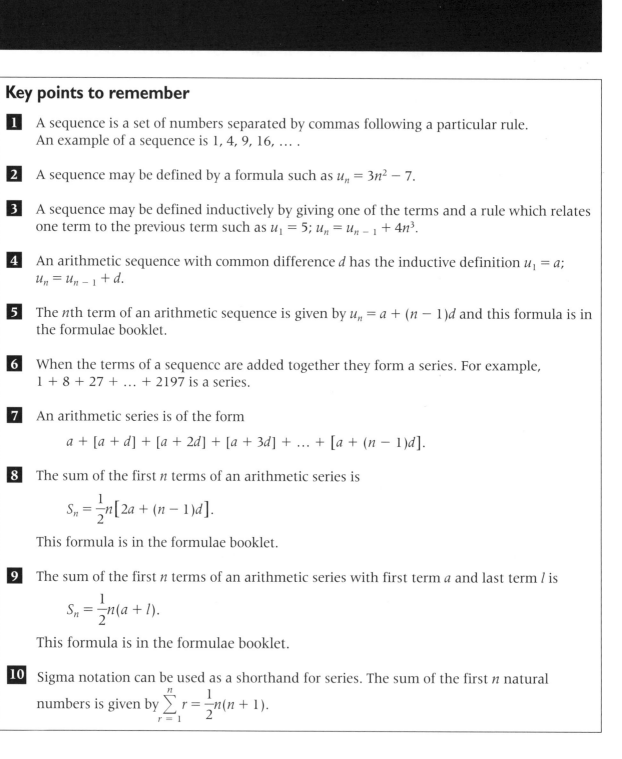

Key points to remember

1 A sequence is a set of numbers separated by commas following a particular rule. An example of a sequence is 1, 4, 9, 16, ...

2 A sequence may be defined by a formula such as $u_n = 3n^2 - 7$.

3 A sequence may be defined inductively by giving one of the terms and a rule which relates one term to the previous term such as $u_1 = 5$; $u_n = u_{n-1} + 4n^3$.

4 An arithmetic sequence with common difference d has the inductive definition $u_1 = a$; $u_n = u_{n-1} + d$.

5 The nth term of an arithmetic sequence is given by $u_n = a + (n-1)d$ and this formula is in the formulae booklet.

6 When the terms of a sequence are added together they form a series. For example, $1 + 8 + 27 + \ldots + 2197$ is a series.

7 An arithmetic series is of the form
$$a + [a + d] + [a + 2d] + [a + 3d] + \ldots + [a + (n-1)d].$$

8 The sum of the first n terms of an arithmetic series is
$$S_n = \frac{1}{2}n[2a + (n-1)d].$$
This formula is in the formulae booklet.

9 The sum of the first n terms of an arithmetic series with first term a and last term l is
$$S_n = \frac{1}{2}n(a + l).$$
This formula is in the formulae booklet.

10 Sigma notation can be used as a shorthand for series. The sum of the first n natural numbers is given by $\sum_{r=1}^{n} r = \frac{1}{2}n(n + 1)$.

Worked example 1

The nth term, u_n, of a sequence is given by $u_n = n^2 - 4$.

(a) Find the first three terms of the sequence.

(b) Find a simplified expression, in terms of n, for:

    (i) $u_{2n} - u_n$

    (ii) $u_n - u_{n-1}$.

(a) $u_n = n^2 - 4$

  $\Rightarrow u_1 = 1^2 - 4 = -3, u_2 = 2^2 - 4 = 0, u_3 = 3^2 - 4 = 5$ ◦—

  The first three terms of the sequence are $-3, 0, 5$.

> The first three terms of the sequence are u_1, u_2, and u_3.

> Putting $n = 1, 2$ and 3

(b) $u_n = n^2 - 4$

  (i) $u_{2n} = (2n)^2 - 4 = 4n^2 - 4$ ◦———

    $\Rightarrow u_{2n} - u_n = 4n^2 - 4 - (n^2 - 4)$

    $\Rightarrow u_{2n} - u_n = 3n^2$

  (ii) $u_{n-1} = (n-1)^2 - 4$ ◦———

    $\Rightarrow u_{n-1} = n^2 - 2n + 1 - 4 = n^2 - 2n - 3$

    $\Rightarrow u_n - u_{n-1} = n^2 - 4 - (n^2 - 2n - 3)$

    $\Rightarrow u_n - u_{n-1} = 2n - 1$

> Replacing n by $2n$

> Replacing n by $n - 1$

Worked example 2

A sequence with nth term, u_n, is defined by

$$u_{n+1} = \alpha u_n + 24, \quad u_1 = 10$$

where α is a constant.

(a) Given that $u_2 = 30$, find the value of α.

(b) The limiting value of u_n as n tends to infinity is L. By forming and solving an equation for L, find the value of L.

8

(a) $u_{n+1} = \alpha u_n + 24 \Rightarrow u_2 = \alpha u_1 + 24$ ◦———

      $\Rightarrow 30 = 10\alpha + 24$

      $\Rightarrow 6 = 10\alpha$

      $\Rightarrow \alpha = 0.6$

> Putting $n = 1$

(b) $u_{n+1} = 0.6u_n + 24$

  As n tends to infinity,

  $u_n \to L$ and so must $u_{n+1} \to L \Rightarrow L = 0.6L + 24$ ◦———

               $\Rightarrow 0.4L = 24$

               $\Rightarrow L = \dfrac{24}{0.4} = 60$

> Replacing both u_n and u_{n+1} by L to form the equation in L

> You can check this by finding later terms in the sequence and noting that as n gets large, the terms approach the value 60 ... eg $u_3 = 42$, $u_7 = 57.6672$, $u_{14} = 59.93 \ldots$, $u_{15} = 59.96 \ldots$

Worked example 3

The ninth term of an arithmetic series is 59 and the fourth term is 24.

(a) Show that the common difference is 7 and find the first term of the series.

(b) The last term of the series is 661. Find the sum of all the terms of the series.

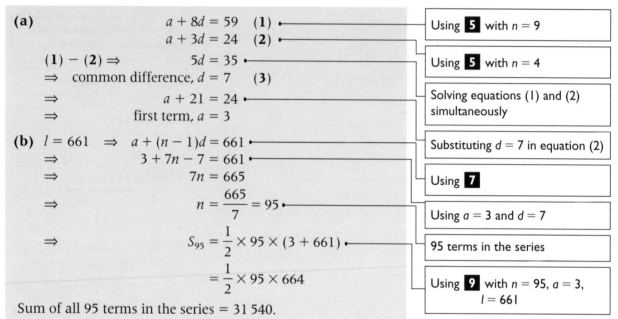

(a)

$$a + 8d = 59 \quad (1)$$
$$a + 3d = 24 \quad (2)$$

$(1) - (2) \Rightarrow \qquad 5d = 35$
$\Rightarrow$ common difference, $d = 7 \quad (3)$
$\Rightarrow \qquad a + 21 = 24$
$\Rightarrow \qquad$ first term, $a = 3$

(b) $l = 661 \Rightarrow a + (n-1)d = 661$
$\Rightarrow \qquad 3 + 7n - 7 = 661$
$\Rightarrow \qquad 7n = 665$
$\Rightarrow \qquad n = \dfrac{665}{7} = 95$

$\Rightarrow \qquad S_{95} = \dfrac{1}{2} \times 95 \times (3 + 661)$

$$= \dfrac{1}{2} \times 95 \times 664$$

Sum of all 95 terms in the series = 31 540.

Using **5** with $n = 9$

Using **5** with $n = 4$

Solving equations (1) and (2) simultaneously

Substituting $d = 7$ in equation (2)

Using **7**

Using $a = 3$ and $d = 7$

95 terms in the series

Using **9** with $n = 95$, $a = 3$, $l = 661$

Worked example 4

(a) Show that the sum of the first n terms of the arithmetic series
$$1 + 5 + 9 + 13 + \ldots$$
is $n(2n - 1)$.

(b) Find the sum of the first 50 terms of the series.

(c) The nth term of the series is u_n. Find the value of $\displaystyle\sum_{n=51}^{100} u_n$.

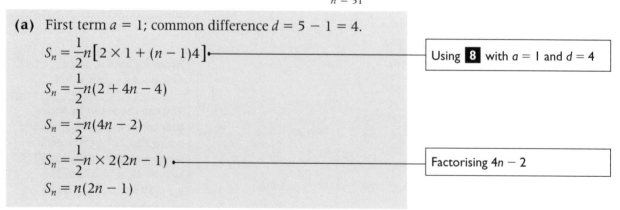

(a) First term $a = 1$; common difference $d = 5 - 1 = 4$.

$$S_n = \dfrac{1}{2}n[2 \times 1 + (n-1)4]$$

$$S_n = \dfrac{1}{2}n(2 + 4n - 4)$$

$$S_n = \dfrac{1}{2}n(4n - 2)$$

$$S_n = \dfrac{1}{2}n \times 2(2n - 1)$$

$$S_n = n(2n - 1)$$

Using **8** with $a = 1$ and $d = 4$

Factorising $4n - 2$

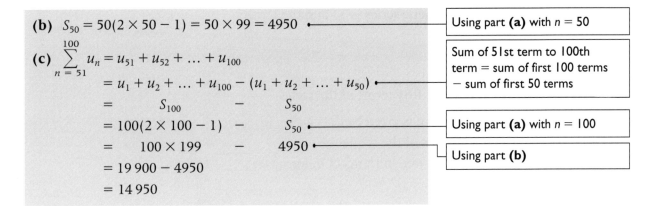

(b) $S_{50} = 50(2 \times 50 - 1) = 50 \times 99 = 4950$ ← Using part **(a)** with $n = 50$

(c) $\displaystyle\sum_{n=51}^{100} u_n = u_{51} + u_{52} + \ldots + u_{100}$

Sum of 51st term to 100th term = sum of first 100 terms − sum of first 50 terms

$\quad = u_1 + u_2 + \ldots + u_{100} - (u_1 + u_2 + \ldots + u_{50})$

$\quad = \qquad S_{100} \qquad - \qquad S_{50}$

$\quad = 100(2 \times 100 - 1) \quad - \qquad S_{50}$ ← Using part **(a)** with $n = 100$

$\quad = \qquad 100 \times 199 \qquad - \quad 4950$ ← Using part **(b)**

$\quad = 19\,900 - 4950$

$\quad = 14\,950$

REVISION EXERCISE 8

1 Find the sum of the first 18 terms of the arithmetic series
$3 + 6 + 9 + \ldots + 54$.

2 A sequence with nth term, u_n, is defined by $u_n = 2u_{n-1} - 3$,
$u_1 = 4$. Find the values of u_2, u_3 and u_4.

3 An arithmetic sequence $u_1, u_2, u_3, \ldots$ has nth term u_n, where
$u_n = 146 - 5n$.

  (a) Write down the values of u_1, u_2 and u_3.

  (b) **(i)** Show that the sequence has exactly 29 positive
        terms.

      (ii) Find the sum of the positive terms of the sequence.

4 For the arithmetic series $9 + 13 + 17 + 21 + \ldots$, find:

  (a) the 100th term,

  (b) the sum of the first 20 terms.

5 The nth term of a sequence is u_n, where $u_n = n(n - 1)$.

  (a) Find the first three terms of the sequence.

  (b) Find an expression in terms of k for the $(k + 1)$th term
     of the sequence.

6 The first term of an arithmetic sequence is 12 and the
common difference is 3.

  (a) Find the ninth term of the sequence.

  (b) The last term of the sequence is 132. Calculate the
     number of terms in the sequence.

7 A sequence with nth term, u_n, is defined by

$$u_{n+1} = ku_n + 7, \quad u_1 = 20$$

where k is a constant.

  (a) Given that $u_2 = 11$, find the value of k.

  (b) The limiting value of u_n as n tends to infinity is L. By
     forming and solving an equation for L, find the value of L.

8

8 The second term of an arithmetic series is 13 and the tenth term of the series is 69.

  (a) Find the common difference.

  (b) Find the first term.

  (c) Find the sum of the first fifty terms of the series.

9 The nth term, u_n, of a sequence is given by $u_n = 4n^2 - 3$.

  (a) Find the first three terms of the sequence.

  (b) Find a simplified expression, in terms of n, for:

  (i) $u_{2n} - u_n$ **(ii)** $u_{n+1} - u_n$.

10 The first term of an arithmetic series is 8. The ninth term of the series is twice the thirteenth term.

  (a) Find the common difference of the series.

  (b) The sum of the first k terms of the series is 0. Find the value of k.

11 The nth term of an arithmetic sequence is u_n, where $u_n = 30 - 4n$.

  (a) Find the values of u_1, u_2 and u_3.

  (b) State the common difference of the sequence.

  (c) **(i)** Evaluate $\displaystyle\sum_{n=1}^{20} u_n$.

  (ii) Evaluate $\displaystyle\sum_{n=1}^{50} u_n$.

  (d) Hence evaluate $\displaystyle\sum_{n=21}^{50} u_n$.

12 The sum of the first 20 terms of an arithmetic series is 400. The tenth term is 19.

  (a) Find the common difference of the series.

  (b) The last term of the series is 95. Calculate the number of terms in the sequence.

13 A sequence with nth term, u_n, is defined by

$$u_{n+1} = a + bu_n, \quad u_1 = 48$$

where a and b are constants. It is given that $u_2 = 20$ and $u_3 = 27$.

  (a) Find the values of the constants a and b.

  (b) The limiting value of u_n as n tends to infinity is L. By forming and solving an equation for L, find the value of L.

14 The nth term of an arithmetic sequence is u_n, where $u_n = 2n - 17$.

  (a) Find the values of u_3 and u_4.

  (b) Find the value of $\displaystyle\sum_{n=5}^{20} u_n$.

15 An arithmetic series has 240 terms. The first term of the series is 64 and the sum of all the terms in the series is 60. Find the last term in the series.

16 The nth term of an arithmetic sequence is u_n, where $u_n = 4n - 2$.

   (a) Find an expression, in terms of n for
   (i) the $(2n)$th term,
   (ii) $u_{2n} - u_n$.

   (b) Find the value of $\sum_{n=1}^{10} (u_{2n} - u_n)$.

Test yourself	**What to review**
	If your answer is incorrect:

1 The nth term of a sequence is u_n, where $u_n = n^2 + 2$.
   (a) Write down the first three terms of the sequence.
   (b) Find a simplified expression, in terms of n, for
   $\qquad 2u_{n+1} - u_n - u_{n-1}$.
	See p 47 Example 1 or review Advancing Maths for AQA C1C2 pp 317–318.

2 A sequence with nth term, u_n, is defined by
   $\qquad u_{n+1} = K - 0.5u_n, \quad u_1 = 8$
   where K is a constant.
   (a) Given that $u_2 = 2$, find the value of K.
   (b) Find the limiting value of u_n as n tends to infinity.
	See p 47 Example 2 or review Advancing Maths for AQA C1C2 pp 318–321.

3 The kth term of the arithmetic series
   $\qquad 258 + 255 + 252 + \ldots$
   is -9. Find the value of k.
	See p 48 Example 3 or review Advancing Maths for AQA C1C2 pp 322–324.

4 The first term of an arithmetic series is -16 and the sum of the first twenty terms of the series is 250. Find the common difference of the series.	See p 48 Example 3 or review Advancing Maths for AQA C1C2 pp 327–328.

5 An arithmetic series has second term 18 and tenth term 42.
   (a) Find the first term of the series.
   (b) The last term of the series is 168. Find the sum of all the terms of the series.
	See p 48 Example 3 or review Advancing Maths for AQA C1C2 pp 326 and 332–333.

6 The nth term of an arithmetic sequence is u_n, where $u_n = 4n + 7$. Find the value of $\sum_{n=11}^{50} u_n$.	See p 48 Example 4 or review Advancing Maths for AQA C1C2 pp 329–332.

8

Test yourself  ANSWERS

1 (a) 3, 6, 11

(b) $6n + 1$

2 (a) 6

(b) 4

3 90

4 3

5 (a) 15

(b) 4758

6 5160

CHAPTER 9

Radian measure

Key points to remember

1 One radian is the angle subtended at the centre of a circle by an arc whose length is equal to the radius of the circle.

      1 radian is written as 1 rad or as 1^c
      1 rad $\approx 57°$

2 Converting between degrees and radians use $360° = 2\pi$ rads.

Hence $x° = x \times \dfrac{2\pi}{360}$ rads.

3 The length l of an arc of a circle is given by $l = r\theta$, where r is the radius and θ is the angle, in radians, subtended by the arc at the centre of the circle.

4 The area A of a sector of a circle is given by $A = \dfrac{1}{2}r^2\theta$, where r is the radius and θ is the angle, in radians, subtended by the arc at the centre of the circle.

Worked example 1

In quadrilateral $ABCD$, angle $A = \dfrac{\pi}{3}$ radians, angle $B = \dfrac{\pi}{2}$ radians,

angle $C = \dfrac{3\pi}{4}$ radians and angle $D = \theta$ radians.

(a) Find θ in terms of π.

(b) Find the size of angle D in degrees.

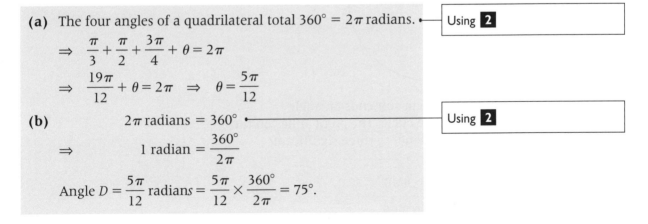

(a) The four angles of a quadrilateral total $360° = 2\pi$ radians.   ——— Using **2**

$$\Rightarrow \quad \frac{\pi}{3} + \frac{\pi}{2} + \frac{3\pi}{4} + \theta = 2\pi$$

$$\Rightarrow \quad \frac{19\pi}{12} + \theta = 2\pi \quad \Rightarrow \quad \theta = \frac{5\pi}{12}$$

(b) 2π radians $= 360°$ ——— Using **2**

$$\Rightarrow \quad \quad \text{1 radian} = \frac{360°}{2\pi}$$

Angle $D = \dfrac{5\pi}{12}$ radian$s = \dfrac{5\pi}{12} \times \dfrac{360°}{2\pi} = 75°$.

9

Worked example 2

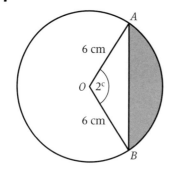

The diagram shows a chord AB of a circle, centre O, radius 6 cm. Angle $AOB = 2$ radians. Calculate the perimeter of the shaded segment, giving your answer to the nearest millimetre.

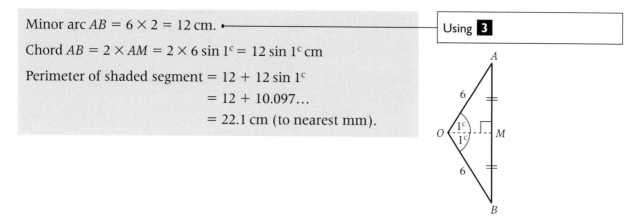

Minor arc $AB = 6 \times 2 = 12$ cm.

Using **3**

Chord $AB = 2 \times AM = 2 \times 6 \sin 1^c = 12 \sin 1^c$ cm

$$\begin{aligned}
\text{Perimeter of shaded segment} &= 12 + 12 \sin 1^c \\
&= 12 + 10.097... \\
&= 22.1 \text{ cm (to nearest mm).}
\end{aligned}$$

Worked example 3

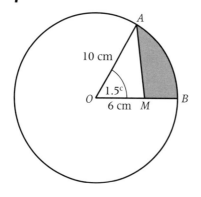

An arc AB of a circle of radius 10 cm subtends an angle 1.5 radians at the centre, O, of the circle. The point M lies on the radius OB such that $OM = 6$ cm. Find, to three significant figures:

(a) the area of the shaded region AMB,

(b) the perimeter of the shaded region.

(a) Area of sector $OAB = \dfrac{1}{2} \times 10^2 \times 1.5 = 75 \text{ cm}^2$.

Using **4**

Area of triangle $OAM = \dfrac{1}{2} \times 10 \times 6 \sin 1.5^c = 30 \sin 1.5^c \text{ cm}^2$.

Using $\dfrac{1}{2}ab \sin C$

Area of shaded region $= 75 - 30 \sin 1.5^c = 45.075 \ldots$
$$= 45.1 \text{ cm}^2$$
(to 3 significant figures)

(b) Arc $AB = 10 \times 1.5 = 15 \text{ cm}$

Using **3**

$$AM^2 = 10^2 + 6^2 - 2 \times 10 \times 6 \cos 1.5^c$$

Using cosine rule

$$AM^2 = 100 + 36 - 120 \cos 1.5^c$$
$$AM^2 = 136 - 8.488 \ldots$$
$$\Rightarrow \quad AM = \sqrt{127.51} \ldots = 11.292 \ldots$$

$MB = OB - OM = 10 - 6 = 4 \text{ cm}$

Perimeter of shaded region $=$ arc $AB + AM + MB$
$$= 15 + 11.29\ldots + 4$$
$$= 30.29\ldots = 30.3 \text{ cm}$$
(to 3 significant figures).

Worked example 4

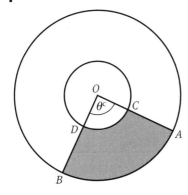

The diagram shows two concentric circles with centre O. The radius of the larger circle is three times the radius of the smaller circle. Points A and B lie on the circumference of the larger circle and angle $AOB = \theta$ radians. Radii OA and OB intersect the smaller circle at the points C and D respectively. The region bounded by the arc AB, the arc CD and the lines AC and BD is shaded.

(a) Express the area of the shaded region in terms of r and θ.

(b) Given that the minor arc AB, is the same length as the major arc CD, show that $\theta = \dfrac{\pi}{2}$.

9

(a) Let the radius of the smaller circle be $r \Rightarrow$ radius of the larger circle $= 3r$.

Area of shaded region = area of sector AOB
$$- \text{ area of sector } COD$$
$$= \frac{1}{2} \times (3r)^2 \times \theta - \frac{1}{2} \times r^2 \times \theta \bullet$$

Using **4** for each circle

$$= 4r^2\theta$$

(b) Minor arc $AB = (3r) \times \theta = 3r\theta \bullet$

Using **3** for larger circle

Major arc subtends an angle $(2\pi - \theta)$ radians at O. $\bullet$
$\Rightarrow$ Major arc $CD = r \times (2\pi - \theta) = r(2\pi - \theta)$. $\bullet$

Reflex angle COD

Minor arc AB = major arc $CD \Rightarrow 3r\theta = r(2\pi - \theta)$
$$\Rightarrow 3r\theta = 2\pi r - r\theta$$
$$\Rightarrow 4r\theta = 2\pi r$$
$$\Rightarrow \quad \theta = \frac{2\pi r}{4r} = \frac{\pi}{2}$$

Using **3** for smaller circle or use $2\pi r$ − minor arc CD

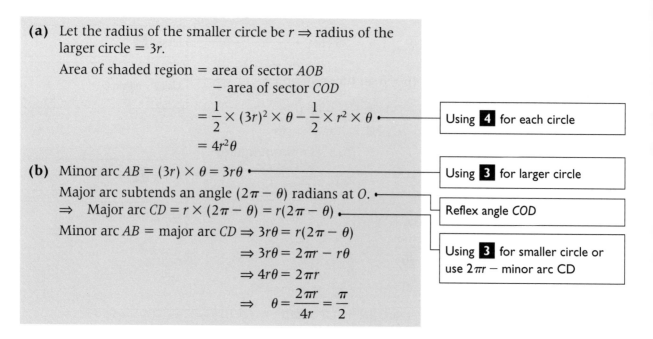

REVISION EXERCISE 9

1 In triangle ABC, angle $A = \dfrac{\pi}{5}$ radians, angle $B = \dfrac{2\pi}{3}$ radians and angle $C = \theta$ radians.

   (a) Find θ in terms of π.

   (b) Find the size of angle C in degrees.

2 The diagram shows a sector of a circle with centre O and radius r cm.

The arc AB subtends an angle θ at O.
(Where appropriate give answers to three significant figures.)

   (a) Find the length of the arc AB in the following cases:
      (i) $\theta = 0.6$ radians, $r = 5$
      (ii) $\theta = 80°$, $r = 9$.

   (b) Find the area of sector AOB in the following cases:
      (i) $\theta = 1.2$ radians, $r = 5$
      (ii) $\theta = 72°$, $r = 10$.

   (c) Given that $r = 6$ and $\theta = 0.6$ radians calculate the length of the **chord** AB.

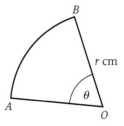

3 The diagram shows a sector of a circle with centre O and radius 8 cm. The length of arc PQ is 9.6 cm.

   (a) Find, in radians, the size of angle POQ.

   (b) Calculate the area of sector POQ.

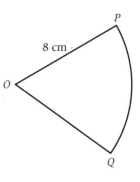

4 The diagram shows a sector of a circle with centre O and radius 6 cm. The area of sector $AOB = 9$ cm^2 and angle $AOB = \theta$ radians.

(a) Find the value of θ.

(b) Find the perimeter of sector AOB.

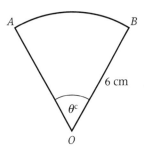

5 The area of a sector of a circle of radius 8 cm is 25.6 cm^2. Find the arc length of the sector.

6 The perimeter of a sector of a circle of radius r cm and angle θ radians is the same as the perimeter of an equilateral triangle of side r cm. Show that $\theta = 1$.

7 The diagram shows a shaded segment of a circle with centre O and radius r m. Angle $OCD = \dfrac{\pi}{3}$ radians

(a) Find, in terms of π, the size of angle COD.

(b) Given that the perimeter of the shaded segment is 12.0 m, calculate the value of r, giving your answer to 3 s.f.

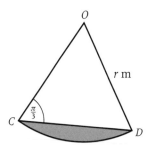

8 The area of a sector of a circle of radius r cm is A cm^2. The perimeter of the sector is P cm. Show that $P = 2r + \dfrac{2A}{r}$.

9

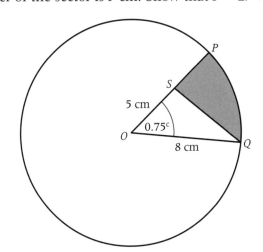

An arc PQ of a circle of radius 8 cm subtends an angle 0.75 radians at the centre, O, of the circle. The point S lies on the radius OP such that $OS = 5$ cm.

(a) Find the area of the sector OPQ.

(b) The region bounded by the arc PQ and the lines PS and SQ is shaded. Find, giving your answers to 3 s.f.
(i) the area of the shaded region,
(ii) the perimeter of the shaded region.

9

Test yourself	**What to review**

If your answer is incorrect:

1 *ABC* is an isosceles triangle with angle $A = \dfrac{7\pi}{12}$ radians.

Angle $B = \theta$ radians.

(a) Find θ in terms of π.

(b) Find the size of angle *C* in degrees.

See p 53 Example 1 or
review Advancing Maths for
AQA C1C2 pp 337–339.

2

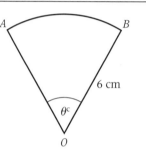

The diagram shows a sector of a circle with centre *O* and radius 6 cm. The perimeter of sector *AOB* = 17.4 cm and angle *AOB* = θ radians. Find the value of θ.

See p 54 Example 2 or
review Advancing Maths for
AQA C1C2 pp 340–342.

3

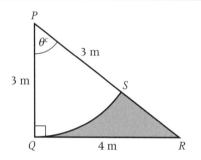

The diagram shows a triangle *PQR* with *PQ* = 3 m, *QR* = 4 m and angle *QPR* = θ radians.

PQ is perpendicular to *QR*.

(a) Show that $\theta = 0.927$ correct to 3 significant figures.

(b) The point *S* lies on *PR* such that *PS* = 3 m and *PQS* is a sector of a circle with centre *P* and radius 3 m. Find, giving your answers to 3 significant figures.
   (i) the area of sector *PQS*,
   (ii) the area of the shaded region bounded by the arc *QS* and the lines *QR* and *SR*.

See pp 54–55 Examples 3
and 4 or review Advancing
Maths for AQA C1C2
pp 342–345.

Test yourself ANSWERS

1 (a) $\dfrac{5\pi}{24}$ (b) $37.5°$

2 0.9

3 (b) (i) 4.17 m² (to 3 significant figures)   (ii) 1.83 m² (to 3

CHAPTER 10

Further trigonometry with radians

Key points to remember

1

θ in degrees	θ in radians
0	0
30	$\dfrac{\pi}{6}$
45	$\dfrac{\pi}{4}$
60	$\dfrac{\pi}{3}$
90	$\dfrac{\pi}{2}$
180	π
360	2π

2 If the interval for θ is given in terms of π you are to assume that θ is measured in radians. In such cases all solutions to the trigonometrical equation must be given in radians.

Worked example 1

(a) The diagram shows a curve with equation $y = \sin x$ for $-2\pi \leqslant x \leqslant 2\pi$ and a horizontal line with equation $y = k$.

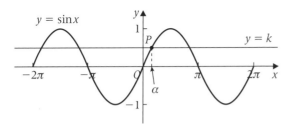

The line intersects the curve at four points, including the point P, where $x = \alpha$, $0 < \alpha < \dfrac{\pi}{2}$.

Find, in terms of π and α, the x-coordinates of the other three points where the line intersects the curve.

(b) Use part (a) to solve the equation $\sin x = \sin \dfrac{4\pi}{9}$, in the interval $-2\pi \leqslant x \leqslant 2\pi$, giving your answers in terms of π.

10

(a)

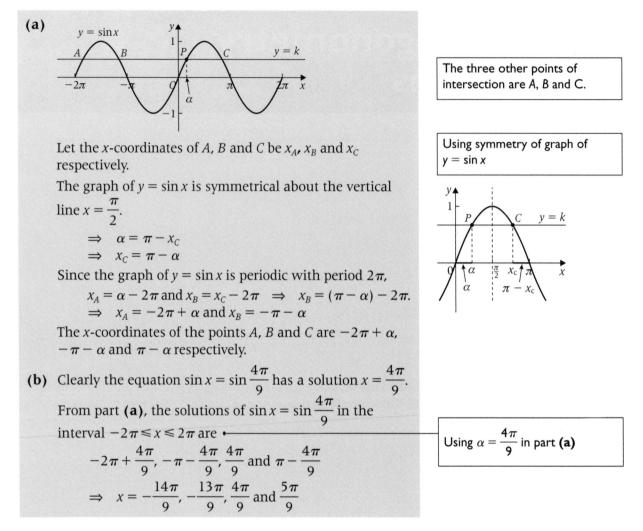

Let the x-coordinates of A, B and C be x_A, x_B and x_C respectively.

The graph of $y = \sin x$ is symmetrical about the vertical line $x = \dfrac{\pi}{2}$.

> Using symmetry of graph of $y = \sin x$

$$\Rightarrow \quad \alpha = \pi - x_C$$
$$\Rightarrow \quad x_C = \pi - \alpha$$

Since the graph of $y = \sin x$ is periodic with period 2π,

$$x_A = \alpha - 2\pi \text{ and } x_B = x_C - 2\pi \quad \Rightarrow \quad x_B = (\pi - \alpha) - 2\pi.$$
$$\Rightarrow \quad x_A = -2\pi + \alpha \text{ and } x_B = -\pi - \alpha$$

The x-coordinates of the points A, B and C are $-2\pi + \alpha$, $-\pi - \alpha$ and $\pi - \alpha$ respectively.

(b) Clearly the equation $\sin x = \sin \dfrac{4\pi}{9}$ has a solution $x = \dfrac{4\pi}{9}$.

From part **(a)**, the solutions of $\sin x = \sin \dfrac{4\pi}{9}$ in the interval $-2\pi \leqslant x \leqslant 2\pi$ are

> Using $\alpha = \dfrac{4\pi}{9}$ in part **(a)**

$$-2\pi + \frac{4\pi}{9}, \; -\pi - \frac{4\pi}{9}, \; \frac{4\pi}{9} \text{ and } \pi - \frac{4\pi}{9}$$

$$\Rightarrow \quad x = -\frac{14\pi}{9}, \; -\frac{13\pi}{9}, \; \frac{4\pi}{9} \text{ and } \frac{5\pi}{9}$$

Worked example 2

Solve the equation $\tan 2\theta = -0.654$ in the interval $-\pi < \theta < \pi$. Give your answers to 3 s.f.

Let $u = 2\theta \quad \Rightarrow \tan u = -0.654.$

> Writing in the form $\tan u = k$

$-\pi < \theta < \pi \quad \Rightarrow -2\pi < 2\theta < 2\pi \quad \Rightarrow -2\pi < u < 2\pi$
or $-6.283\ldots < u < 6.283\ldots$

> Finding interval for u

$\tan u = -0.654 \Rightarrow u = \tan^{-1}(-0.654)$
$$\Rightarrow u = -0.5791\ldots \text{ radians}$$

> Using **2**; set calculator in radian mode.

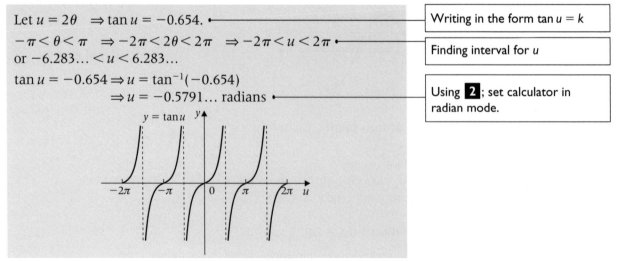

$y = \tan u$ is periodic with period π.

$\Rightarrow \quad u = -0.5791\ldots, -0.5791\ldots \pm \pi, -0.5791\ldots \pm 2\pi, \ldots$

$\Rightarrow \quad u = -0.5791\ldots, -3.7207\ldots, 2.5624\ldots, 5.7040\ldots$

$\Rightarrow \quad 2\theta = -3.7207\ldots, -0.5791\ldots, 2.5624\ldots, 5.7040\ldots$

$\Rightarrow \quad \theta = -1.86^c, -0.290^c, 1.28^c, 2.85^c$

> Choosing values of u in interval $-2\pi < u < 2\pi$

> Giving answers to the degree of accuracy asked for

Worked example 3

(a) Find the angle θ, in the interval $0° < \theta < 90°$, which satisfies the equation $\cos\theta = \dfrac{1}{2}$.

(b) Solve the equation $\cos\left(x + \dfrac{\pi}{4}\right) = \dfrac{1}{2}$ in the interval $0 < x < 4\pi$. Give your answers in the form $k\pi$.

(a) $\cos\theta = \dfrac{1}{2} \Rightarrow \theta = \cos^{-1} 0.5$

$\Rightarrow \theta = 60°$

> Interval $0° < \theta < 90°$ is in degrees

(b) Let $u = x + \dfrac{\pi}{4} \Rightarrow \cos u = \dfrac{1}{2}$

> Writing in the form $\cos u = k$

$\Rightarrow u = \dfrac{\pi}{3}$

> Using part **(a)** and **I**

$0 < x < 4\pi \Rightarrow \dfrac{\pi}{4} < x + \dfrac{\pi}{4} < 4\pi + \dfrac{\pi}{4} \Rightarrow \dfrac{\pi}{4} < u < \dfrac{17\pi}{4}$

> Finding interval for u

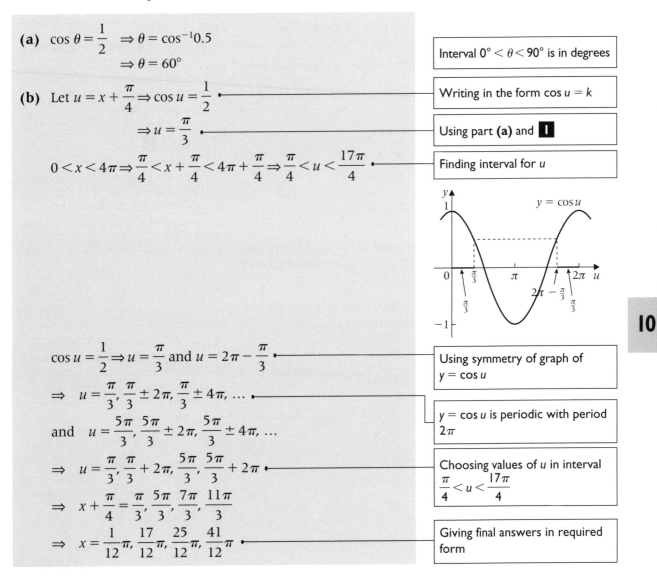

$\cos u = \dfrac{1}{2} \Rightarrow u = \dfrac{\pi}{3}$ and $u = 2\pi - \dfrac{\pi}{3}$

> Using symmetry of graph of $y = \cos u$

$\Rightarrow u = \dfrac{\pi}{3}, \dfrac{\pi}{3} \pm 2\pi, \dfrac{\pi}{3} \pm 4\pi, \ldots$

> $y = \cos u$ is periodic with period 2π

and $u = \dfrac{5\pi}{3}, \dfrac{5\pi}{3} \pm 2\pi, \dfrac{5\pi}{3} \pm 4\pi, \ldots$

$\Rightarrow u = \dfrac{\pi}{3}, \dfrac{\pi}{3} + 2\pi, \dfrac{5\pi}{3}, \dfrac{5\pi}{3} + 2\pi$

> Choosing values of u in interval $\dfrac{\pi}{4} < u < \dfrac{17\pi}{4}$

$\Rightarrow x + \dfrac{\pi}{4} = \dfrac{\pi}{3}, \dfrac{5\pi}{3}, \dfrac{7\pi}{3}, \dfrac{11\pi}{3}$

$\Rightarrow x = \dfrac{1}{12}\pi, \dfrac{17}{12}\pi, \dfrac{25}{12}\pi, \dfrac{41}{12}\pi$

> Giving final answers in required form

10

Worked example 4

It is given that $(\sin x - 2\cos x)(\sin x + \cos x) = 0$.

(a) Find the two possible values of $\tan x$.

(b) Hence solve the equation $(\sin x - 2\cos x)(\sin x + \cos x) = 0$, in the interval $-2\pi \leqslant x \leqslant 2\pi$, giving your answers to 3 s.f.

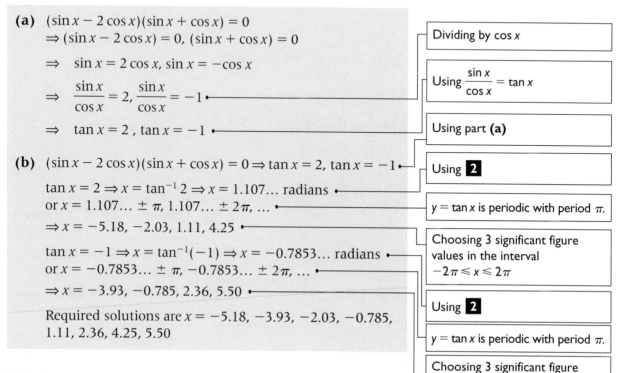

(a) $(\sin x - 2\cos x)(\sin x + \cos x) = 0$
 $\Rightarrow (\sin x - 2\cos x) = 0, (\sin x + \cos x) = 0$ ⎤ Dividing by $\cos x$

 $\Rightarrow \quad \sin x = 2\cos x, \sin x = -\cos x$

 $\Rightarrow \quad \dfrac{\sin x}{\cos x} = 2, \dfrac{\sin x}{\cos x} = -1$ • Using $\dfrac{\sin x}{\cos x} = \tan x$

 $\Rightarrow \quad \tan x = 2, \tan x = -1$ • Using part **(a)**

(b) $(\sin x - 2\cos x)(\sin x + \cos x) = 0 \Rightarrow \tan x = 2, \tan x = -1$ • Using **2**

 $\tan x = 2 \Rightarrow x = \tan^{-1} 2 \Rightarrow x = 1.107\ldots$ radians •
 or $x = 1.107\ldots \pm \pi, 1.107\ldots \pm 2\pi, \ldots$ • — $y = \tan x$ is periodic with period π.
 $\Rightarrow x = -5.18, -2.03, 1.11, 4.25$ •

Choosing 3 significant figure values in the interval $-2\pi \leqslant x \leqslant 2\pi$

 $\tan x = -1 \Rightarrow x = \tan^{-1}(-1) \Rightarrow x = -0.7853\ldots$ radians •
 or $x = -0.7853\ldots \pm \pi, -0.7853\ldots \pm 2\pi, \ldots$ •
 $\Rightarrow x = -3.93, -0.785, 2.36, 5.50$ • — Using **2**

Required solutions are $x = -5.18, -3.93, -2.03, -0.785,$
1.11, 2.36, 4.25, 5.50

$y = \tan x$ is periodic with period π.

Choosing 3 significant figure values in the interval $-2\pi \leqslant x \leqslant 2\pi$

Worked example 5

(a) Use an appropriate trigonometrical identity to show that the equation $6\cos^2 x + 11\sin x - 9 = 0$ can be written as $(2\sin x - 3)(3\sin x - 1) = 0$.

(b) Hence find the values of x for which $6\cos^2 x + 11\sin x - 9 = 0$ and $0 \leqslant x \leqslant 2\pi$, giving your answers to 3 s.f.

(a) $6\cos^2 x + 11\sin x - 9 = 0$
 $\Rightarrow \quad 6(1 - \sin^2 x) + 11\sin x - 9 = 0$ • Using identity $\cos^2 x + \sin^2 x = 1$
 $\Rightarrow \quad 6\sin^2 x - 11\sin x + 3 = 0$
 $\Rightarrow \quad (2\sin x - 3)(3\sin x - 1) = 0$ • Factorising quadratic in $\sin x$

(b) $6\cos^2 x + 11\sin x - 9 = 0$
 $\Rightarrow 2\sin x - 3 = 0$ or $3\sin x - 1 = 0$
 $\Rightarrow \quad \sin x = \dfrac{3}{2}$ or $\sin x = \dfrac{1}{3}$

Since $-1 \leqslant \sin x \leqslant 1$, $\sin x \neq \dfrac{3}{2}$.

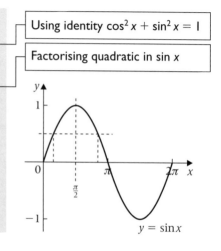

$y = \sin x$

$\sin x = \frac{1}{3} \Rightarrow x = 0.3398...$ radians

or $x = \pi - 0.3398...^c$

In the interval $0 \leqslant x \leqslant 2\pi$, $x = 0.340$, 2.80 to 3 significant figures

Using **2**

Graph of $y = \sin x$ is symmetrical about $x = \dfrac{\pi}{2}$

REVISION EXERCISE 10

1 For each of the following equations, find the solution that lies in the interval $0 < \theta < \dfrac{\pi}{2}$, giving your answer to 3 s.f.

   (a) $\sin \theta = 0.866$ (b) $\cos \theta = 0.678$

   (c) $\tan \theta = 1$ (d) $\cos 2\theta = 0.5$

   (e) $\tan 2\theta = 4$

2 Solve the equation $\sin(x + 1.2) = -0.56$ in the interval $0 < x < 2\pi$, giving your answers to 3 s.f.

3 Solve the equation $\cos \dfrac{1}{2}x = -0.48$ in the interval $-\pi < x < 2\pi$. Give your answer to 2 d.p.

4 (a) Sketch the graph of the curve $y = \cos x$ for $-2\pi \leqslant x \leqslant 2\pi$.

   (b) The horizontal line $y = k$ intersects the curve $y = \cos x$ at a point P, where $x = \alpha$, $0 < \alpha < \dfrac{\pi}{2}$. Find, in terms of π and α, the x-coordinates of other three points where the line intersects the curve in the interval $-2\pi \leqslant x \leqslant 2\pi$.

   (c) Solve the equation $\cos x = \cos \dfrac{2\pi}{5}$, in the interval $-2\pi \leqslant x \leqslant 2\pi$, giving your answers in terms of π.

5 (a) Sketch the graph of the curve $y = \tan x$ for $-2\pi \leqslant x \leqslant 2\pi$.

   (b) The horizontal line $y = k$ intersects the curve $y = \tan x$ at a point Q, where $x = \alpha$, $0 < \alpha < \dfrac{\pi}{2}$. Find, in terms of π and α, the x-coordinates of other points where the line intersects the curve in the interval $-2\pi \leqslant x \leqslant 2\pi$.

   (c) Solve the equation $\tan x = \tan \dfrac{\pi}{6}$, in the interval $-2\pi \leqslant x \leqslant 2\pi$, giving your answers in terms of π.

6 Solve the equation $\sin 2x = \sin \dfrac{2\pi}{5}$ for $0 \leqslant x \leqslant 2\pi$, giving your answers in terms of π.

7 Solve the equation $\cos(x + 1.3) = 0.345$ giving all solutions to three significant figures in the interval $-\pi < x < \pi$.

10

8 For a particular value of k, one solution of the equation $\sin x = k$ is $x = 0.42$. Find, to two significant figures, the solutions of $\sin x = k$ that lie in the interval $\pi \leqslant x \leqslant 4\pi$.

9 Solve the equation $\tan\left(x - \dfrac{\pi}{3}\right) = \tan\dfrac{\pi}{4}$, in the interval $-2\pi \leqslant x \leqslant 2\pi$. Leave your answers in terms of π.

10 Solve the equation $\cos 2x = \cos\left(-\dfrac{\pi}{3}\right)$, in the interval $-\pi \leqslant x \leqslant \pi$.

11 It is given that $4 \sin x = 7 \cos x$.

   (a) Write down the value of $\tan x$.

   (b) Hence solve the equation $4 \sin x = 7 \cos x$ in the interval $0 < x < 2\pi$, giving your answers to 3 s.f.

12 Solve the equation $(2 \cos x - \sin x)(\cos x + \sin x) = 0$ in the interval $-\pi < x < \pi$, giving your answer to 2 d.p.

13 It is given that θ satisfies the equation $3 \sin^2 \theta = 3 - \cos \theta$.

   (a) Use an appropriate trigonometrical identity to show that $3 \cos^2 \theta - \cos \theta = 0$.

   (b) By solving this quadratic equation find the two possible values of $\cos \theta$ and hence find all the possible values of θ in the interval $0 < \theta < 2\pi$, giving your answers to 3 s.f.

14 Solve the equation $\tan x(2 \sin x - \cos x) = 0$, in the interval $-\pi \leqslant x \leqslant \pi$, giving your answers to 3 s.f. where appropriate.

15 (a) Show that the equation $2 \tan \theta(1 + \sin \theta) = -5 \cos \theta$ can be written as $3 \sin^2 \theta - 2 \sin \theta - 5 = 0$.

   (b) Hence show that the equation $2 \tan \theta(1 + \sin \theta) = -5 \cos \theta$ has only one solution in the interval $-\pi \leqslant \theta \leqslant \pi$ and state the value of this solution to 3 s.f.

16 (a) Solve the equation $(2 \sin \theta - 1)(2 \cos \theta - 1) = 0$, in the interval $0° \leqslant \theta \leqslant 360°$.

   (b) Hence solve the equation $(2 \sin 2x - 1)(2 \cos 2x - 1) = 0$ in the interval $0 \leqslant x \leqslant \pi$, giving your answers in terms of π.

Test yourself	**What to review**
	If your answer is incorrect:
1 Solve the equation $\sin x = 0.75$ in the interval $0 < x < 2\pi$. Give your answers to 2 decimal places.	See p 61 Example 3 or review Advancing Maths for AQA C1C2 pp 349–351.
2 Solve the equation $\cos 2x = 0.63$ in the interval $-\pi < x < \pi$. Give your answers to three 3 significant figures.	See p 60 Example 2 or review Advancing Maths for AQA C1C2 pp 351–352.
3 Solve the equation $\tan\left(\theta - \dfrac{\pi}{3}\right) = \tan\dfrac{\pi}{4}$ in the interval $0 < \theta < 2\pi$, leaving your answers in terms of π.	See p 59 Example 1 or review Advancing Maths for AQA C1C2 pp 352–353.
4 Solve the equation $\cos x(6\sin x - \cos x) = 0$ in the interval $-\pi < x < \pi$. Give your answers to 3 decimal places.	See p 62 Example 4 or review Advancing Maths for AQA C1C2 pp 354–356.
5 (a) Show that the equation $4(\sin^2 x + \cos x) = 1$ can be written in the form $(2\cos x - 3)(2\cos x + 1) = 0$. **(b)** Hence solve the equation $4(\sin^2 x + \cos x) = 1$ in the interval $0 < x < 2\pi$. Give your answers to 3 significant figures.	See p 62 Example 5 or review Advancing Maths for AQA C1C2 pp 354–356.

Test yourself ANSWERS

5 $2.09, 4.19$

4 $-2.976, -1.571, 0.165, 1.571$

3 $\dfrac{7\pi}{12}, \dfrac{19\pi}{12}$

2 $\mp 0.445, \pm 2.70$

1 $0.85, 2.29$

10

CHAPTER 11

Exponentials and logarithms

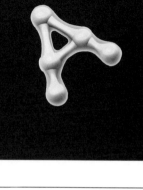

Key points to remember

1 An **exponent** is another name for an index or power.
Functions such as f where $f(x) = 2^x$ are called **exponential functions** because the variable x appears in the exponent.

2 For all values of the positive constant a, the curve $y = a^x$ will always pass through the point $(0, 1)$.

3 The logarithm of N to base a is written as $\log_a N$.

4 $N = a^x \Leftrightarrow x = \log_a N$

5 $\log_a a = 1$ and $\log_a 1 = 0$

6 The laws of logarithms are:

$$\log_a x + \log_a y = \log_a(xy), \quad \text{[Law 1]}$$
$$\log_a x - \log_a y = \log_a\left(\frac{x}{y}\right), \quad \text{[Law 2]}$$
$$k \log_a x = \log_a(x^k). \quad \text{[Law 3]}$$

7 $\log_{10} x$ is sometimes written as $\lg x$.

8 The solution to the equation $a^x = b$ is $x = \dfrac{\log_{10} b}{\log_{10} a}$, where logarithms are taken to base 10.

9 In general, to find the value of $\log_a b$:
(i) let $x = \log_a b \Rightarrow a^x = b$,
(ii) take logarithms to base 10 of both sides $\Rightarrow \log_{10} a^x = \log_{10} b$,
(iii) use Law 3 $\Rightarrow x \log_{10} a = \log_{10} b$,
(iv) rearrange $\Rightarrow x = \dfrac{\log_{10} b}{\log_{10} a}$ so

$$\log_a b = \frac{\log_{10} b}{\log_{10} a} \text{ then evaluate the right-hand side using a calculator.}$$

Worked example 1

Find the value of:

(a) $\log_4 4$ **(b)** $\log_4 1$ **(c)** $\log_4 2$ **(d)** $\log_4 8$.

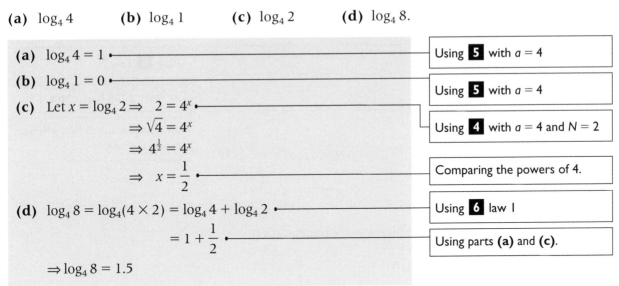

(a) $\log_4 4 = 1$ Using **5** with $a = 4$

(b) $\log_4 1 = 0$ Using **5** with $a = 4$

(c) Let $x = \log_4 2 \Rightarrow 2 = 4^x$ Using **4** with $a = 4$ and $N = 2$

$\Rightarrow \sqrt{4} = 4^x$

$\Rightarrow 4^{\frac{1}{2}} = 4^x$

$\Rightarrow x = \dfrac{1}{2}$ Comparing the powers of 4.

(d) $\log_4 8 = \log_4(4 \times 2) = \log_4 4 + \log_4 2$ Using **6** law 1

$= 1 + \dfrac{1}{2}$ Using parts **(a)** and **(c)**.

$\Rightarrow \log_4 8 = 1.5$

Worked example 2

(a) Describe the geometrical transformation by which the curve with equation $y = 5^x - 2$ can be obtained from the curve with equation $y = 5^x$.

(b) Sketch the graph of the curve $y = 5^x - 2$.

(c) The point $(k, 5)$ lies on the curve $y = 5^x - 2$. Use logarithms to find the value of k, giving your answer to 4 d.p.

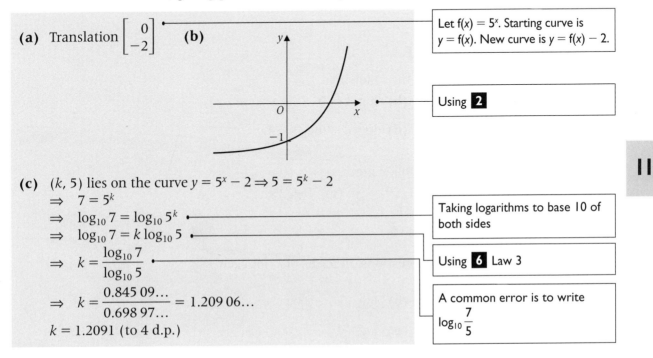

(a) Translation $\begin{bmatrix} 0 \\ -2 \end{bmatrix}$ **(b)**   Let $f(x) = 5^x$. Starting curve is $y = f(x)$. New curve is $y = f(x) - 2$.

Using **2**

(c) $(k, 5)$ lies on the curve $y = 5^x - 2 \Rightarrow 5 = 5^k - 2$

$\Rightarrow 7 = 5^k$

$\Rightarrow \log_{10} 7 = \log_{10} 5^k$ Taking logarithms to base 10 of both sides

$\Rightarrow \log_{10} 7 = k \log_{10} 5$

$\Rightarrow k = \dfrac{\log_{10} 7}{\log_{10} 5}$ Using **6** Law 3

$\Rightarrow k = \dfrac{0.845\,09\ldots}{0.698\,97\ldots} = 1.209\,06\ldots$ A common error is to write $\log_{10}\dfrac{7}{5}$

$k = 1.2091$ (to 4 d.p.)

11

Worked example 3

Given that $\log_a x = 2\log_a 8 - \log_a 4$, where a is a positive constant, show that $x = 16$.

$$\Rightarrow \quad \log_a x = \log_a 8^2 - \log_a 4$$

Using **6** Law 3

$$\Rightarrow \quad \log_a x = \log_a\left(\frac{8^2}{4}\right)$$

Using **6** Law 2

$$\Rightarrow \quad x = \frac{8^2}{4} = \frac{64}{4}$$

Comparing the logarithms

$$\Rightarrow \quad x = 16$$

Always show full details when an answer is given.

Worked example 4

Find the two integer values of n which satisfy the equation

$$2\log_a n = \log_a(n-2) + \frac{1}{2}\log_a 81.$$

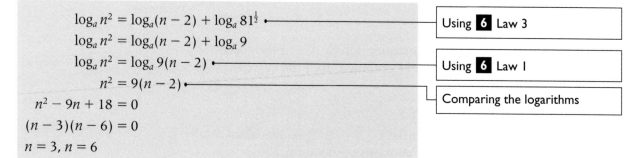

$$\log_a n^2 = \log_a(n-2) + \log_a 81^{\frac{1}{2}}$$

Using **6** Law 3

$$\log_a n^2 = \log_a(n-2) + \log_a 9$$

$$\log_a n^2 = \log_a 9(n-2)$$

Using **6** Law 1

$$n^2 = 9(n-2)$$

Comparing the logarithms

$$n^2 - 9n + 18 = 0$$

$$(n-3)(n-6) = 0$$

$$n = 3, n = 6$$

REVISION EXERCISE 11

1 Find the value of k in each of the following:

   (a) $\log_2 k = 1$ **(b)** $\log_3 k = 0$

   (c) $\log_{16} k = \dfrac{1}{2}$ **(d)** $\log_2 k - \log_2 3 = 4.$

2 Find the value of each of the following:

   (a) $\log_5 5$ **(b)** $\log_5 1$ **(c)** $\log_5 25$

   (d) $\log_5 \dfrac{1}{25}$ **(e)** $\log_5 \dfrac{1}{5\sqrt{5}}.$

3 Given that $\log_a 4 = x$, find expressions, in terms of x, for each of the following:

   (a) $\log_a 16$ **(b)** $\log_a 2$

   (c) $\log_a 32$ **(d)** $\log_a \dfrac{1}{4}.$

4 Given that $\log_x a = 2$ and $\log_x b = 3$, express ab in terms of x.

5 (a) Sketch the graph of $y = 4^{-x}$.

(b) Describe the geometrical transformation by which the curve with equation $y = 4^{-2x}$ can be obtained from the curve with equation $y = 4^{-x}$.

(c) The point $(k, 5)$ lies on the curve $y = 4^{-2x}$. Use logarithms to find the value of k, giving your answer to 4 d.p.

6 Given that $\log_a m - \log_a n^2 = 2 \log_a a + \log_a m^3$, where a, m and n are positive constants. Show that $mn = \dfrac{1}{a}$.

7 Write each of the following as a single logarithm:

(a) $\log_a 10 + \log_a 5$ **(b)** $\log_a 10 - \log_a 5$

(c) $3 \log_a a + \log_a 4$ **(d)** $2 \log_a 4 + 1$.

8 Solve the equation $5^x = 31$, giving your answer to 5 s.f.

9 Solve the equation $\log_3(27x) = 5$.

10 The line $y = 5$ meets the curve $y = 2^{2x-1}$ at the point P. Show that the x-coordinate of P is $\dfrac{1}{\log_{10} 4}$.

11 (a) Using the substitution $y = 4^x$ show that the equation $16^x - 4^{x+1} + 3 = 0$ can be written as $(y - 1)(y - 3) = 0$.

(b) Hence show that the equation $16^x - 4^{x+1} + 3 = 0$ has a solution $x = 0$ and find the other solution, giving your answer to 4 d.p.

12 A curve has equation $y = 2 \log_{10} x - \log_{10} 2$, where $x > 0$. The point $A(4, p)$ and the point $B(6, q)$ lie on the curve.

(a) Show that $p = \log_{10} 8$.

(b) Express q in the form $\log_{10} k$.

(c) Hence show that the gradient of the chord AB is $\log_{10} \dfrac{3}{2}$.

11

Test yourself	**What to review**

<table>
<tr><td></td><td>If your answer is incorrect:</td></tr>
</table>

1 (a) Sketch the graph of $y = 2^x$.

(b) State the geometrical transformation which maps the graph of $y = 2^x$ onto the graph of $y = 2^{\frac{x}{2}}$

See p 67 Example 2 or review Advancing Maths for AQA C1C2 pp 359–361.

---

2 Evaluate the following:

(a) $\log_2 32$ **(b)** $\dfrac{\log_3 27}{\log_4 2}$ **(c)** $\log_4 \sqrt{2}$.

See p 67 Example 1 or review Advancing Maths for AQA C1C2 pp 361–362.

---

3 Given that $4 \log_a p - \log_a q = 2$, where a, p and q are positive numbers, express a in terms of p and q.

See p 68 Example 3 or review Advancing Maths for AQA C1C2 pp 363–364.

---

4 It is given that $2^x = 3$ and $3^y = 4$.

(a) Use logarithms to find the value of x, giving your answer to 6 significant figures.

(b) Show that $xy = 2$.

Review Advancing Maths for AQA C1C2 pp 365–368.

---

5 Solve the equation $\log_4 x = 3 \log_4 3 - \log_4 9 + \log_4 1 - 1$.

See p 68 Example 4 or review Advancing Maths for AQA C1C2 pp 363–364.

Test yourself **ANSWERS**

5 $x = \dfrac{3}{4}$

4 (a) 1.584 96

3 $a = \dfrac{p^2}{\sqrt{q}}$

2 (a) 5 **(b)** 6 **(c)** $\dfrac{1}{4}$

(b) Stretch scale factor 2 parallel to the x-axis

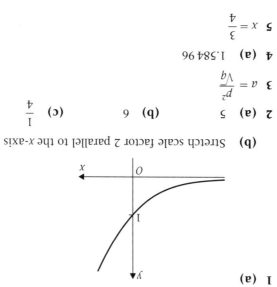

1 (a)

CHAPTER 12
Geometric series

Key points to remember

1 A geometric series is of the form

$$a + ar + ar^2 + ar^3 + \ldots$$

with first term a and common ratio r.

2 The nth term of a geometric series is given by ar^{n-1}.
This formula is in the formulae booklet.

3 The sum of the first n terms of a geometric series is

$$S_n = \frac{a(1 - r^n)}{(1 - r)}.$$

This formula is in the formulae booklet.

4 A geometric series with common ratio r converges when $|r| < 1$.

5 A convergent geometric series has sum to infinity

$$S_\infty = \frac{a}{(1 - r)}.$$

This formula is in the formulae booklet.

Worked example 1

The second term of a geometric series is 1024 and the eighth term is 16. Given that all the terms in the series are positive, find the common ratio and the first term of the series.

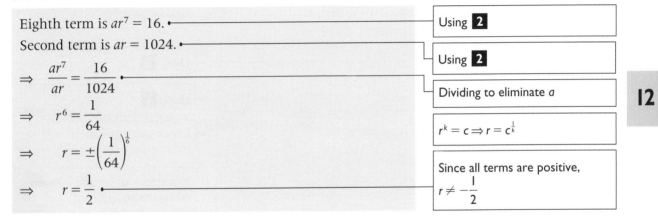

Eighth term is $ar^7 = 16$. ──────────── Using **2**

Second term is $ar = 1024$. ──────────── Using **2**

$\Rightarrow \quad \dfrac{ar^7}{ar} = \dfrac{16}{1024}$ ──────────── Dividing to eliminate a

$\Rightarrow \quad r^6 = \dfrac{1}{64}$

$\Rightarrow \quad r = \pm\left(\dfrac{1}{64}\right)^{\frac{1}{6}}$ ──────────── $r^k = c \Rightarrow r = c^{\frac{1}{k}}$

$\Rightarrow \quad r = \dfrac{1}{2}$ ──────────── Since all terms are positive, $r \neq -\dfrac{1}{2}$

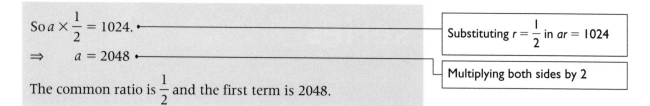

So $a \times \dfrac{1}{2} = 1024$.

⟶ Substituting $r = \dfrac{1}{2}$ in $ar = 1024$

$\Rightarrow \quad a = 2048$

⟶ Multiplying both sides by 2

The common ratio is $\dfrac{1}{2}$ and the first term is 2048.

Worked example 2

Find the sum of the first thirty terms of the geometric series
$72 - 48 + 32 - \ldots$, giving your answer to 4 d.p.

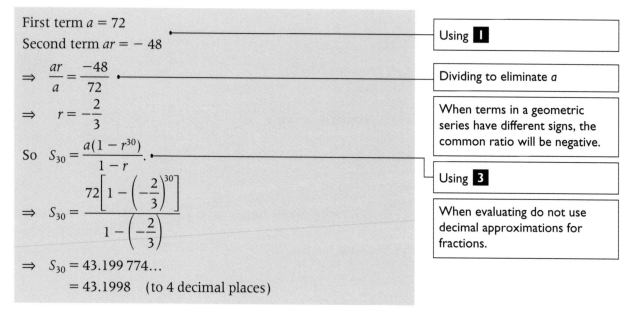

First term $a = 72$
Second term $ar = -48$

⟶ Using **1**

$\Rightarrow \quad \dfrac{ar}{a} = \dfrac{-48}{72}$

⟶ Dividing to eliminate a

$\Rightarrow \quad r = -\dfrac{2}{3}$

When terms in a geometric series have different signs, the common ratio will be negative.

So $\quad S_{30} = \dfrac{a(1 - r^{30})}{1 - r}$

⟶ Using **3**

$\Rightarrow \quad S_{30} = \dfrac{72\left[1 - \left(-\dfrac{2}{3}\right)^{30}\right]}{1 - \left(-\dfrac{2}{3}\right)}$

When evaluating do not use decimal approximations for fractions.

$\Rightarrow \quad S_{30} = 43.199\,774\ldots$
$\qquad\quad = 43.1998 \quad$ (to 4 decimal places)

Worked example 3

The third term of a geometric series is 2187 and the fifth term is
243.

(a) Find the two possible values for the common ratio.

(b) Explain why both possible geometric series converge.

(c) Find the sum to infinity for each of the series.

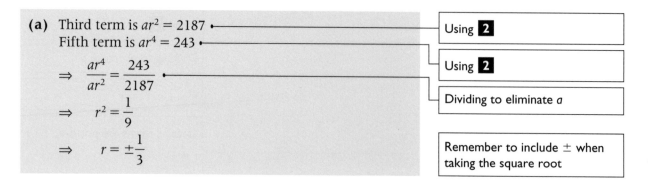

(a) Third term is $ar^2 = 2187$

⟶ Using **2**

Fifth term is $ar^4 = 243$

⟶ Using **2**

$\Rightarrow \quad \dfrac{ar^4}{ar^2} = \dfrac{243}{2187}$

⟶ Dividing to eliminate a

$\Rightarrow \quad r^2 = \dfrac{1}{9}$

$\Rightarrow \quad r = \pm\dfrac{1}{3}$

Remember to include $\pm$ when taking the square root

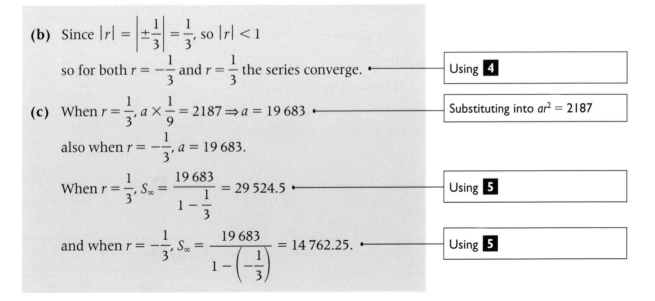

(b) Since $|r| = \left|\pm\dfrac{1}{3}\right| = \dfrac{1}{3}$, so $|r| < 1$

so for both $r = -\dfrac{1}{3}$ and $r = \dfrac{1}{3}$ the series converge. — Using **4**

(c) When $r = \dfrac{1}{3}$, $a \times \dfrac{1}{9} = 2187 \Rightarrow a = 19\,683$ — Substituting into $ar^2 = 2187$

also when $r = -\dfrac{1}{3}$, $a = 19\,683$.

When $r = \dfrac{1}{3}$, $S_\infty = \dfrac{19\,683}{1 - \dfrac{1}{3}} = 29\,524.5$ — Using **5**

and when $r = -\dfrac{1}{3}$, $S_\infty = \dfrac{19\,683}{1 - \left(-\dfrac{1}{3}\right)} = 14\,762.25$. — Using **5**

Worked example 4

A convergent geometric series has first term 6250 and its sum to infinity is 15 625. Find the fifth term of the series.

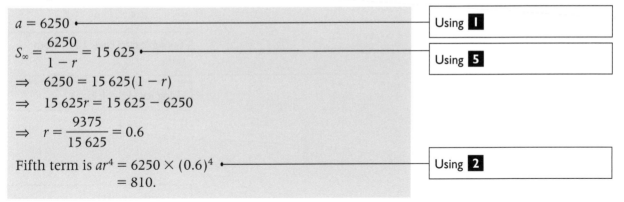

$a = 6250$ — Using **1**

$S_\infty = \dfrac{6250}{1 - r} = 15\,625$ — Using **5**

$\Rightarrow \quad 6250 = 15\,625(1 - r)$

$\Rightarrow \quad 15\,625r = 15\,625 - 6250$

$\Rightarrow \quad r = \dfrac{9375}{15\,625} = 0.6$

Fifth term is $ar^4 = 6250 \times (0.6)^4$ — Using **2**

$\qquad\qquad = 810$.

Worked example 5

Find the set of values of x for which the geometric series
$1024 + 768x + 576x^2 + 432x^3 + \dots$ converges.

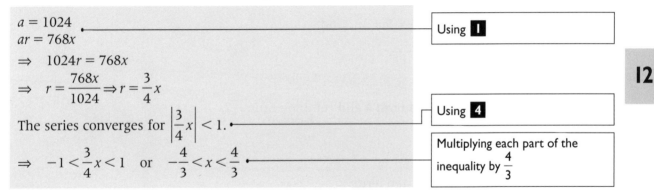

$a = 1024$ — Using **1**
$ar = 768x$

$\Rightarrow \quad 1024r = 768x$

$\Rightarrow \quad r = \dfrac{768x}{1024} \Rightarrow r = \dfrac{3}{4}x$

The series converges for $\left|\dfrac{3}{4}x\right| < 1$. — Using **4**

$\Rightarrow \quad -1 < \dfrac{3}{4}x < 1 \quad$ or $\quad -\dfrac{4}{3} < x < \dfrac{4}{3}$ — Multiplying each part of the inequality by $\dfrac{4}{3}$

12

Worked example 6

A geometric series begins $50 + 40 + 32 + 25.6 + \dots$

(a) Find the common ratio of the series.

(b) Show that the sum of the first 27 terms is approximately 249.4.

(c) Find the least value of n such that the nth term is less than 0.5.

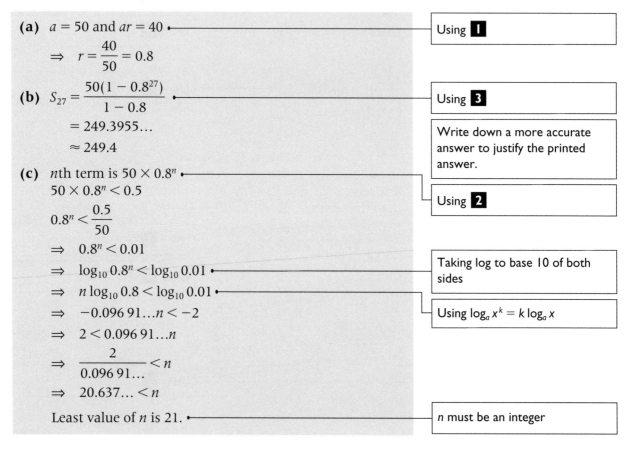

(a) $a = 50$ and $ar = 40$ Using **1**

$\Rightarrow \quad r = \dfrac{40}{50} = 0.8$

(b) $S_{27} = \dfrac{50(1 - 0.8^{27})}{1 - 0.8}$ Using **3**

$\qquad\quad = 249.3955\dots$ Write down a more accurate answer to justify the printed answer.

$\qquad\quad \approx 249.4$

(c) nth term is 50×0.8^n Using **2**

$\quad 50 \times 0.8^n < 0.5$

$\quad 0.8^n < \dfrac{0.5}{50}$

$\Rightarrow \quad 0.8^n < 0.01$

$\Rightarrow \quad \log_{10} 0.8^n < \log_{10} 0.01$ Taking log to base 10 of both sides

$\Rightarrow \quad n \log_{10} 0.8 < \log_{10} 0.01$ Using $\log_a x^k = k \log_a x$

$\Rightarrow \quad -0.096\,91\dots n < -2$

$\Rightarrow \quad 2 < 0.096\,91\dots n$

$\Rightarrow \quad \dfrac{2}{0.096\,91\dots} < n$

$\Rightarrow \quad 20.637\dots < n$

Least value of n is 21. n must be an integer

REVISION EXERCISE 12

1 For the geometric series $4 + 6 + 9 + \dots$, find:

   (a) the common ratio,

   (b) the seventh term.

2 A geometric series has second term 4 and common ratio -2.

   (a) Find the first term.

   (b) Find the tenth term.

   (c) Find the sum of the first ten terms.

3 The second term of a geometric series is 8 and the fifth term is 27.

   (a) Find the common ratio.

   (b) Find the first term.

   (c) Calculate the sum of the first nine terms, giving your answer to 1 d.p.

4 A geometric series is given by $3 + k + 48 + \dots$.
   Find the possible values of the second term, k.

5 The common ratio of a geometric series is $\dfrac{5}{2}$ and the third term is 25.

   (a) Find the first term.

   (b) Calculate the sum of the first seven terms.

6 A geometric series has second term 10 and third term 8. Find the sum of the terms from the fourth term to the twelfth term inclusive, giving your answer to 3 d.p.

7 A convergent geometric series has first term 25 and second term 20. Find the sum to infinity of the series.

8 A geometric series begins $40 + 20 + 10 + 5 + \dots$.

   (a) Explain why the series converges.

   (b) Calculate the sum to infinity of the series.

9 The sum to infinity of a convergent geometric series is 200. The first term of the series is 140.

   (a) Find the common ratio of the series.

   (b) Calculate the sum of the first ten terms of the series, giving your answer to 4 d.p.

10 The common ratio of a convergent geometric series, with sum to infinity equal to 7, is the same as the first term of the series.

   (a) Find the first term of the series.

   (b) The nth term of the series is u_n. Show that
   $$\log_{10}\left(\frac{u_{2k}}{u_k}\right) = k(\log_{10} 7 - \log_{10} 8).$$

11 Find the values of x for which the geometric series
   $$1 + (6 + 4x) + (6 + 4x)^2 + (6 + 4x)^3 + \dots$$
   converges.

12 Find the least number of terms of the geometric series $48 + 60 + 75 + \dots$ required to give a sum greater than 38 400.

13 **(a)** Find the greatest number of terms in the geometric series $81 + 54 + 36 + \dots$ which are greater than 0.01.

   (b) Find the sum to infinity of the series.

12

14 The sum to infinity of a convergent geometric series is $9a$, where a is the first term of the series.

   (a) Show that the common ratio of the series is $\dfrac{8}{9}$.

   (b) The third term of the series is 640.

      (i) Find the first term of the series.

      (ii) Show that the sum of the first seven terms is approximately 4094.

15 Evaluate

   (a) $\displaystyle\sum_{n=1}^{3}\left(\dfrac{1}{2}\right)^n$ **(b)** $\displaystyle\sum_{n=1}^{\infty}\left(\dfrac{1}{2}\right)^n$ **(c)** $\displaystyle\sum_{n=4}^{\infty}\left(\dfrac{1}{2}\right)^n$.

16 The first three terms of a geometric series are the second, fourth and eighth terms of an arithmetic series that has common difference 4.

   (a) Find the first term of the geometric series.

   (b) The sum of the first n terms of the geometric series is S_n.

      (i) Show that $S_{60} = 2^{63} - 8$.

      (ii) Show that $S_{2n} - S_{2n-1} = 2^{2n+2}$.

Test yourself	What to review
	If your answer is incorrect:

1 The third term of a geometric series is 54 and the sixth term is -128.
  (a) Find the common ratio.
  (b) Find the first term.
	See p 71 Example 1 or review Advancing Maths for AQA C1C2 pp 373–374.

2 Find the sum of the first fifteen terms of the geometric series $12 + 18 + 27 + \ldots$, giving your answer to 1 decimal place.	See p 72 Example 2 or review Advancing Maths for AQA C1C2 pp 375–376.
3 Find the set of values of x for which the geometric series $36 + 18(1 - 5x) + 9(1 - 5x)^2 + \ldots$ is convergent.	See p 73 Example 5 or review Advancing Maths for AQA C1C2 pp 376–377.

4 The sum to infinity of a convergent geometric series is seven times the first term of the series.
  (a) Find the common ratio.
  (b) Given that the third term is 180, find the first term.
	See p 73 Example 4 or review Advancing Maths for AQA C1C2 pp 376–378.

5 Find the least number of terms of the geometric series $2 + 3 + 4.5 + \ldots$ required to give a sum greater than four million.	Review Advancing Maths for AQA C1C2 pp 379–381.

Test yourself **ANSWERS**

1 (a) $-\dfrac{3}{4}$ **(b)** 30.375

2 10 485.5

3 $-\dfrac{1}{5} > x > \dfrac{3}{5}$

4 (a) $\dfrac{6}{7}$ **(b)** 245

5 35

12

Exam style practice paper

Answer **all** questions.
Time allowed: 1 hour 30 minutes

---

1 Evaluate $\int_1^2 \frac{1}{x^2}\, dx$.   (3 marks)

2 The diagram shows a chord AB of a circle of centre O.
The minor arc AB is 30 cm and angle AOB is 2.5 radians.

   (a) Show that the radius of the circle is 12 cm.   (2 marks)

   (b) Calculate the area of triangle OAB to the nearest cm².   (2 marks)

   (c) Calculate the area of the shaded segment to the nearest cm².   (3 marks)

   (d) Calculate the perimeter of the shaded segment to 3 significant figures.   (3 marks)

3 The nth term, u_n, of a sequence is defined by $u_{n+1} = ku_n + 5$, where k is a constant. It is given that $u_1 = 15$ and $u_2 = 11$.

   (a) Find the value of k.   (2 marks)

   (b) Find the value of u_3 and u_4.   (2 marks)

   (c) The limiting value of u_n as n tends to infinity is L.
      (i) Write down an equation for L.   (1 mark)
      (ii) Find the value of L.   (2 marks)

4 (a) The fourth term of a geometric series is 162 and the seventh term of the series is 48.
      (i) Find the common ratio of the series.   (3 marks)
      (ii) Find the first term of the series.   (2 marks)

   (b) Evaluate $\sum_{n=1}^{\infty} \left(\frac{1}{4}\right)^n$.   (3 marks)

5 (a) Write down the value of $\log_a a^4$.   (1 mark)

   (b) Given that $\log_a z = 3\log_a x - 2\log_a y - 4$, find z in terms of x, y and a.   (4 marks)

6 The diagram shows the graph of the curve with equation

$y = 4x - \dfrac{4}{3}x\sqrt{x}.$

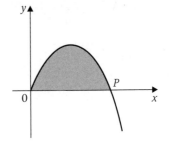

   (a) The curve crosses the x-axis at the point P. Verify that the x-coordinate of P is 9. (1 mark)

   (b) Find $\dfrac{dy}{dx}$. (4 marks)

   (c) Find the equation of the normal to the curve at the point P, giving your answer in the form $ay - x + b = 0$. (5 marks)

   (d) Find the area of the shaded region bounded by the curve and the x-axis. (4 marks)

7 (a) Sketch the graph of $y = \cos x$ for $-2\pi \leqslant x \leqslant 2\pi$. (3 marks)

   (b) One solution of the equation $\cos x = \cos \dfrac{\pi}{7}$ is $x = \dfrac{\pi}{7}$. Find, in terms of π, all the solutions of the equation $\cos x = \cos \dfrac{\pi}{7}$ that lie in the interval $-2\pi \leqslant x \leqslant 2\pi$. (3 marks)

   (c) Describe the single geometrical transformation by which the curve $y = \cos 3x$ can be obtained from the curve $y = \cos x$. (2 marks)

   (d) Solve the equation $\cos 3x = 0.321$, giving all answers in radians in the interval $0 \leqslant x \leqslant \pi$ to 3 significant figures. (4 marks)

8 The curve, C, with equation $y = 2^x - 16$ intersects the x-axis at the point P and intersects the y-axis at the point Q.

   (a) Show that the x-coordinate of P is 4 and state the y-coordinate of Q. (2 marks)

   (b) Describe the single geometrical transformation by which the graph of the curve with equation $y = 2^x - 16$ can be obtained from the graph of $y = 2^x$. (2 marks)

   (c) Sketch the graph of the curve $y = 2^x - 16$. (2 marks)

   (d) (i) Use the trapezium rule with five ordinates (four strips) to find an approximate value for

$$\int_0^4 (2^x - 16)\, dx.$$

(4 marks)

   (ii) Hence find an approximate value for the area of the region bounded by the curve C and the line PQ. (2 marks)

   (e) The line $y = 1$ intersects the curve C at the point R. By using logarithms to solve the equation $2^x = 17$, find the x-coordinate of R, giving your answer to 3 decimal places. (4 marks)

Answers

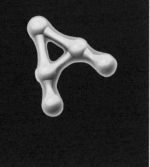

1 $6x^8$

2 $4a^4$

3 $8n^6$

4 $5m^{12}$

5 $6p^4q^2$

6 $3a$

7 $2m^8n^5$

8 a^{-2} or $\dfrac{1}{a^2}$

9 $3x^5$

10 $m^{-6}n^8$ or $\dfrac{n^8}{m^6}$

11 $4x\sqrt{x}$

12 $4x^{\frac{9}{4}}$

13 $x^{\frac{3}{2}}y^{\frac{1}{6}}$

14 $16p^4q^{\frac{1}{6}}$

15 $\dfrac{9y^{\frac{1}{6}}}{2\sqrt{x}}$

16 $x^{\frac{2}{3}} - x^{-\frac{1}{3}}$

17 $x^{\frac{3}{2}} - x^{-\frac{3}{2}}$

18 $x^{\frac{5}{2}} + x^{-\frac{1}{2}} + 2x$

19 (a) (i) $3^{\frac{1}{2}}$ (ii) 3^{3x} (b) $x = -\frac{1}{8}$

20 (a) (i) $4^{\frac{1}{2}}$ (ii) 4^{-2} (iii) $4^{1.5x}$ (b) $x = -8$

21 $x = \frac{1}{2}$

22 (b) $x = 1$

23 $x = -1, \frac{1}{2}$

1 $\dfrac{dy}{dx} = -\dfrac{1}{x^2}$

2 $\dfrac{dy}{dx} = \dfrac{1}{2}x^{-\frac{1}{2}} = \dfrac{1}{2\sqrt{x}}$

3 $\dfrac{dy}{dx} = -\dfrac{1}{2}x^{-\frac{3}{2}} = -\dfrac{1}{2x\sqrt{x}}$

4 $\dfrac{dy}{dx} = -\dfrac{5}{x^2}$

5 $\dfrac{dy}{dx} = 1 - \dfrac{1}{x^3}$

6 $\dfrac{dy}{dx} = 4x + \dfrac{3}{x^3}$

7 $\dfrac{dy}{dx} = 4 + 2x^{-\frac{2}{3}}$

8 $\dfrac{dy}{dx} = \dfrac{2}{\sqrt{x}} - \dfrac{1}{x\sqrt{x}} - 3$

9 $\dfrac{dy}{dx} = \dfrac{1}{3}x^{-\frac{3}{5}} - \dfrac{1}{2}x^{-\frac{4}{3}}$

10 $\dfrac{dy}{dx} = \dfrac{1}{3\sqrt{x}} - \dfrac{3}{2}\sqrt{x}$

11 $\dfrac{dy}{dx} = \dfrac{3}{2}x^{-\frac{1}{4}} + 4$

12 $\dfrac{dy}{dx} = \dfrac{\sqrt{x}}{4} - \dfrac{7}{2}x^2\sqrt{x}$

13 $\dfrac{dy}{dx} = -\dfrac{10}{x^3} - \dfrac{1}{x^2}$

14 $\dfrac{dy}{dx} = 3x^2 + 12\sqrt{x}$

15 $\dfrac{dy}{dx} = -\dfrac{3}{x^4} + \dfrac{1}{x^2}$

16 $\dfrac{dy}{dx} = \dfrac{3}{x^3} - \dfrac{12}{x^4} + \dfrac{8}{x^5}$

17 $\dfrac{dy}{dx} = -3x^{-2} - x^{-\frac{3}{2}} + \dfrac{1}{3}x^{-\frac{5}{3}}$

18 $\dfrac{dy}{dx} = \dfrac{4}{3}x^{-\frac{2}{3}} - \dfrac{2}{3}x^{-\frac{7}{6}} - \dfrac{2}{3}x^{-\frac{5}{3}}$

19 (a) 18   (b) 2

20 $\dfrac{1}{8}, -1$

21 $x = \sqrt{2}$

22 (a) $4x + y = 3$ (b) $\dfrac{9}{8}$

23 (a) e.g. $y - 3 = -2(x - 1)$ (c) $\dfrac{29}{6}$

24 (a) $\dfrac{2}{3}$ (b) $2x + y = 1$ (c) $\dfrac{1}{4}$

25 (a) $(9, 162)$ (b) -7.5, maximum   (c) $x > 9$

26 (b) $66\,m$

Revision exercise 3

1 (a) $y = -x^{-6} + c$ (b) $y = 4x^{\frac{5}{4}} + c$ (c) $y = -2x^{-3} + c = -\dfrac{2}{x^3} + c;$ (d) $y = \dfrac{5}{6}x^{\frac{6}{5}} + c = \dfrac{5}{6}\sqrt[5]{x^6} + c$

2 (a) $2x^{\frac{3}{2}} + x + c$ (b) $3x^3 + \dfrac{1}{2}x^{-1} + c$ (c) $-x^{-1} + 2x^{-3} + c$ (d) $\dfrac{4}{3}x^{\frac{3}{2}} - x^{\frac{1}{2}} + c$

3 (a) $f(x) = -\dfrac{2}{x} + \dfrac{1}{2}x^3 + c$ (b) $f(x) = 3x\sqrt[3]{x} - \sqrt{x} + c$

   (c) $f(x) = 2\sqrt{x} + \dfrac{2}{3}x\sqrt{x} + c$ (d) $f(x) = -\dfrac{3}{x} + x - \dfrac{2}{3}x^3 + c$

4 $y = \dfrac{1}{3}x^3 + 2x - \dfrac{1}{x} - 14$ 5 $y = 2x^2\sqrt{x} - \dfrac{5}{2}x^2 - 27$

6 (a) $y = -\dfrac{1}{x^2} + \dfrac{1}{x} + 3x + 7$ (b) $(1, 10)$

7 (a) 5   (b) $6\dfrac{1}{4}$ (c) $32\dfrac{3}{4}$ (d) $\dfrac{27}{140}$

8 $\dfrac{3}{2}$ 9 (a) $\dfrac{14}{3}$

10 (a) $\dfrac{1}{2}$ (b) $4x + \dfrac{1}{x} + c$ (c) $\dfrac{1}{4}$

11 18.982

12 (a) 2.573   (b)

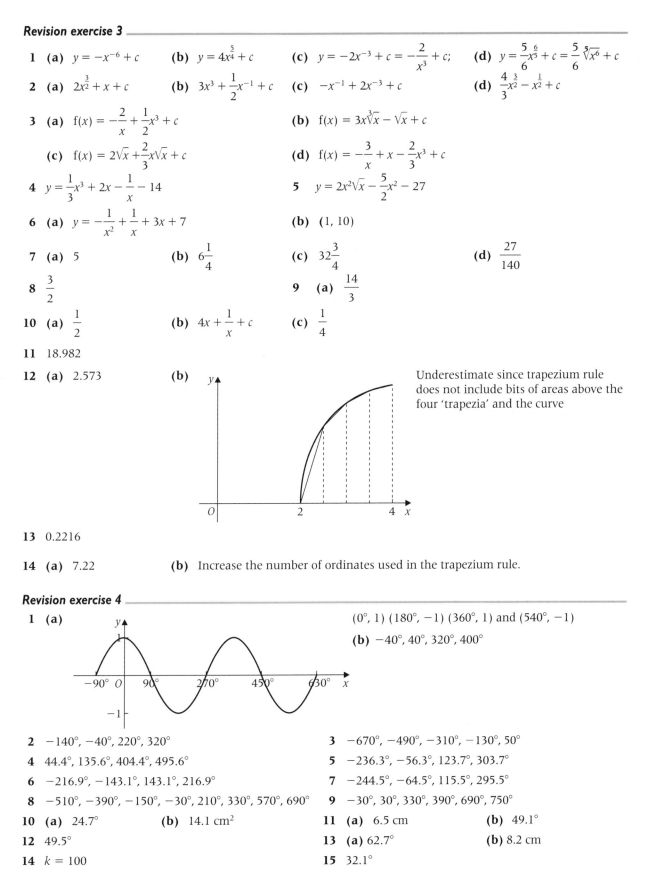

Underestimate since trapezium rule does not include bits of areas above the four 'trapezia' and the curve

13 0.2216

14 (a) 7.22   (b) Increase the number of ordinates used in the trapezium rule.

Revision exercise 4

1 (a) $(0°, 1)$ $(180°, -1)$ $(360°, 1)$ and $(540°, -1)$
          (b) $-40°, 40°, 320°, 400°$

2 $-140°, -40°, 220°, 320°$ 3 $-670°, -490°, -310°, -130°, 50°$
4 $44.4°, 135.6°, 404.4°, 495.6°$ 5 $-236.3°, -56.3°, 123.7°, 303.7°$
6 $-216.9°, -143.1°, 143.1°, 216.9°$ 7 $-244.5°, -64.5°, 115.5°, 295.5°$
8 $-510°, -390°, -150°, -30°, 210°, 330°, 570°, 690°$ 9 $-30°, 30°, 330°, 390°, 690°, 750°$
10 (a) 24.7°   (b) 14.1 cm²   11 (a) 6.5 cm   (b) 49.1°
12 49.5°   13 (a) 62.7°   (b) 8.2 cm
14 $k = 100$ 15 32.1°

Revision exercise 5

1 A translation of $\begin{bmatrix} -28° \\ 0 \end{bmatrix}$

2 **(a)** $\begin{bmatrix} -2 \\ 0 \end{bmatrix}$ **(b)** $y = (x + 2)^2 - 4$

3 $y = 7^{-x}$

4 **(a)** Reflection in the x-axis

(b) Stretch with scale factor $\dfrac{1}{3}$ in the x-direction

(c) Translation $\begin{bmatrix} 1 \\ 4 \end{bmatrix}$

5 **(a)** Translation $\begin{bmatrix} 0 \\ -4 \end{bmatrix}$

(b) Reflection in the x-axis

(c) Stretch with scale factor 3 in the x-direction

(d) Stretch with scale factor 2 in the y-direction

(e) Translation $\begin{bmatrix} 40° \\ 0 \end{bmatrix}$

6 **(a)** $\dfrac{1}{4}$ **(b)** 16

7 Reflection in the y-axis

8 $y = x - 2$

9 **(a)** $y = -3x$ **(b)** $y = 5 - 3x$

10 **(a)** $y = 4x + 8$ **(b)** $y = x + 4$

11 **(a)** Translation $\begin{bmatrix} -30° \\ 0 \end{bmatrix}$

(b) Stretch with scale factor $\dfrac{1}{2}$ in the x-direction

(c) Stretch with scale factor $\dfrac{1}{5}$ in the y-direction

(d) Reflection in the y-axis

12 **(a)** $\begin{bmatrix} -\dfrac{1}{2} \\ 0 \end{bmatrix}$

(b) Stretch with scale factor 2 in the y-direction

13 **(a)**

(b)

(c)

(d)

(e)

(f)

Revision exercise 6

1 $15°, 75°, 135°, 195°, 255°, 315°$

2 $22.2°, 67.8°, 202.2°, 247.8°$

3 $-313.4°, 313.4°$

4 $-332°, -92°, 28°, 268°$

5 $-234.1°, -54.1°, 125.9°$

6 $224.2°, 275.8°$

7 $-360°, -180°, -90°, 0°, 180°, 270°, 360°, 540°, 630°, 720°$

8 $-90°, 0°, 90°, 270°, 360°$

9 $-180°, -45°, 0°, 135°, 180°, 315°, 360°$

10 **(b)** $\pm\dfrac{5}{12}$

11 **(a)** 2

(b) $-99°, -39°, 21°, 81°, 141°$

12 **(b)** **(i)** $-\dfrac{1}{2}, 4$

(ii) $-150°, -30°, 210°, 330°$

14 **(a)** $1 \pm \sqrt{3}$

(b) $137.1°, 222.9°$

15 $-105°, -90°, -15°, 75°, 90°, 165°$

16 $-160.5°, -19.5°, 199.5°, 340.5°$

Revision exercise 7

1 **(a)** 720

(b) 3

(c) 100

2 252

3 **(a)** $1 + 5x + 10x^2 + 10x^3 + 5x^4 + x^5$

(b) $16 - 32x + 24x^2 - 8x^3 + x^4$

(c) $a^3 + 6a^2b + 12ab^2 + 8b^3$

4 **(a)** $16a^4 + 32a^3b + 24a^2b^2 + 8ab^3 + b^4$

(b) $16a^4 - 32a^3b + 24a^2b^2 - 8ab^3 + b^4$

5 **(a)** $1 + 6x + 12x^2 + 8x^3$

(b) $1 - 6x + 12x^2 - 8x^3$

(c) $x = \pm7$

6 **(a)** $1 + 4x + 6x^2 + 4x^3 + x^4$

(b) $17 + 12\sqrt{2}$

7 **(a)** 3003

(b) 3217.5

8 $k = \pm3$

9 $99 - 70\sqrt{2}$

10 **(a)** $1 - 40x + 720x^2 - 7680x^3$

(b) -7140

11 **(a)** $1 + 15x + 90x^2 + 270x^3 + 405x^4 + 243x^5$

(b) 36

(c) 8343

Revision exercise 8

1 513

2 $5, 7, 11$

3 **(a)** $141, 136, 131$

(b) **(ii)** 2059

4 **(a)** 405

(b) 940

5 **(a)** $0, 2, 6$

(b) $k(k + 1)$

6 **(a)** 36

(b) 41

7 **(a)** 0.2

(b) 8.75

8 **(a)** 7

(b) 6

(c) 8875

9 **(a)** $1, 13, 33$

(b) **(i)** $12n^2$

(ii) $8n + 4$

10 **(a)** -0.5

(b) 33

11 (a) 26, 22, 18 **(b)** −4 **(c) (i)** −240 **(ii)** −3600 **(d)** − 3360

12 (a) 2 **(b)** 48

13 (a) $a = 32, b = -0.25$ **(b)** 25.6

14 (a) $u_3 = -11, u_4 = -9$ **(b)** 128

15 −63.5

16 (a) (i) $8n - 2$ **(ii)** $4n$ **(b)** 220

Revision exercise 9

1 (a) $\dfrac{2\pi}{15}$ **(b)** 24°

2 (a) (i) 3 cm **(ii)** 4π cm = 12.6 cm (to 3 significant figures)
(b) (i) 15 cm² **(ii)** 20π cm² = 62.8 cm² (to 3 significant figures)
(c) 3.55 cm (to 3 significant figures)

3 (a) 1.2 rads **(b)** 38.4 cm²

4 (a) 0.5 **(b)** 15 cm

5 6.4 cm

7 (a) $\dfrac{\pi}{3}$ rads **(b)** 5.86

9 (a) 24 cm² **(b) (i)** 10.4 cm² **(ii)** 14.5 cm

Revision exercise 10

1 (a) 1.05 **(b)** 0.826 **(c)** 0.785 **(d)** 0.524 **(e)** 0.663

2 2.54, 4.49

3 4.14

4 (a)

(b) $-\alpha, -2\pi + \alpha, 2\pi - \alpha$ **(c)** $\pm\dfrac{2\pi}{5}, \pm\dfrac{8\pi}{5}$

5 (a)

(b) $-2\pi + \alpha, -\pi + \alpha, \pi + \alpha$
(c) $-\dfrac{11\pi}{6}, -\dfrac{5\pi}{6}, \dfrac{\pi}{6}, \dfrac{7\pi}{6}$

6 $\dfrac{\pi}{5}, \dfrac{3\pi}{10}, \dfrac{6\pi}{5}, \dfrac{13\pi}{10}$

7 $-2.52, -0.0814$

8 $6.7, 9.0$

9 $-\dfrac{17\pi}{12}, -\dfrac{5\pi}{12}, \dfrac{7\pi}{12}, \dfrac{19\pi}{12}$

10 $\pm\dfrac{\pi}{6}, \pm\dfrac{5\pi}{6}$

11 (a) 1.75　　　　　　　　(b) 1.05, 4.19

12 $-2.03, -0.79, 1.11, 2.36$

13 (b) $\cos\theta = 0$ or $\cos\theta = \dfrac{1}{3}$; 1.23, 1.57, 4.71, 5.05

14 $-3.14, -2.68, 0, 0.464, 3.14$

15 (b) -1.57

16 (a) $30°, 60°, 150°, 300°$　　(b) $\dfrac{\pi}{12}, \dfrac{\pi}{6}, \dfrac{5\pi}{12}, \dfrac{5\pi}{6}$

Revision exercise 11

1 (a) 2　　　　　　(b) 1　　　　　　(c) 4　　　　　　(d) 48

2 (a) 1　　　　　　(b) 0　　　　　　(c) 2　　　　　　(d) -2　　　　　(e) $-1\dfrac{1}{2}$

3 (a) $2x$　　　　　(b) $\dfrac{1}{2}x$　　　　(c) $2\dfrac{1}{2}x$　　　　(d) $-x$

4 x^5

5 (a)

 (b) Stretch scale factor 0.5 parallel to the x-axis
 (c) -0.5805 (4 decimal places)

7 (a) $\log_a 50$　　　　(b) $\log_a 2$　　　　(c) $\log_a(4a^3)$　　　　(d) $\log_a 16a$

8 2.1337 (5 significant figures)

9 9

11 (b) 0.7925 (4 decimal places)

12 (b) $q = \log_{10} 18$

Revision exercise 12

1 (a) 1.5　　　　　　(b) 45.5625

2 (a) -2　　　　　(b) 1024　　　　(c) 682

3 (a) $\dfrac{3}{2}$　　　　　(b) $\dfrac{16}{3}$　　　　(c) 399.4

4 ±12   **5 (a)** 4   **(b)** 1624.9375

6 27.705   **7** 125

8 (a) $r = \dfrac{1}{2} \Rightarrow |r| < 1 \Rightarrow$ series converges   **(b)** 80

9 (a) $r = \dfrac{3}{10}$ **(b)** 199.9988

10 (a) $\dfrac{7}{8}$ **11** $-\dfrac{7}{4} < x < -\dfrac{5}{4}$ **12** 24   **13 (a)** 23   **(b)** 243

14 (b) (i) 810

15 (a) $\dfrac{7}{8}$ **(b)** 1   **(c)** $\dfrac{1}{8}$

16 (a) 8

Exam style practice paper

1 $\dfrac{1}{2}$

2 (b) 43 cm²   **(c)** 137 cm²   **(d)** 52.8 cm

3 (a) 0.4   **(b)** $u_3 = 9.4,\ u_4 = 8.76$ **(c) (i)** $L = 0.4L + 5$ **(ii)** $8\dfrac{1}{3}$

4 (a) (i) $\dfrac{2}{3}$ **(ii)** 546.75   **(b)** $\dfrac{1}{3}$

5 (a) 4   **(b)** $z = \dfrac{x^3}{y^2 a^4}$

6 (b) $\dfrac{dy}{dx} = 4 - 2\sqrt{x}$ **(c)** $2y - x + 9 = 0$ **(d)** 32.4

7 (a)   **(b)** $-\dfrac{13\pi}{7}, -\dfrac{\pi}{7}, \dfrac{\pi}{7}, \dfrac{13\pi}{7}$

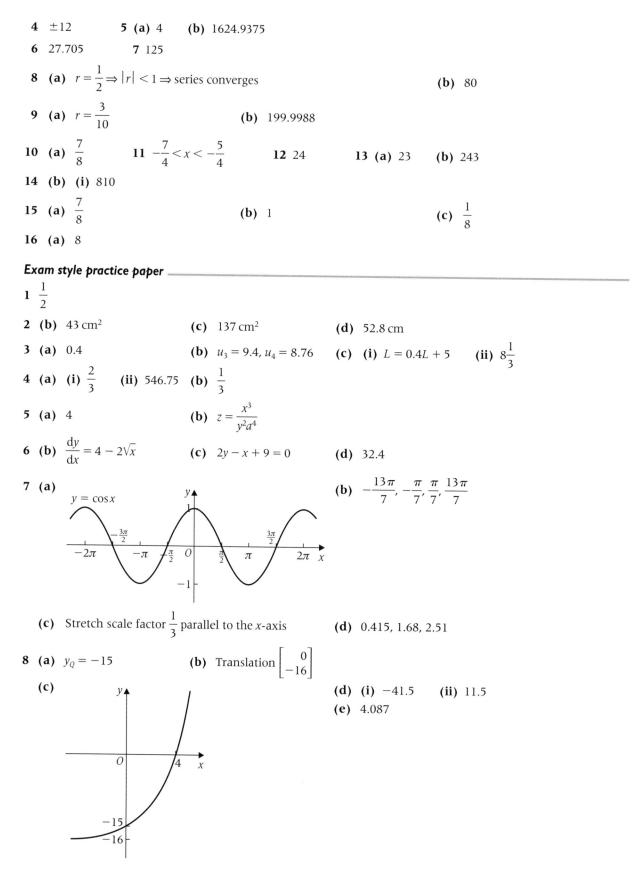

(c) Stretch scale factor $\dfrac{1}{3}$ parallel to the x-axis   **(d)** 0.415, 1.68, 2.51

8 (a) $y_Q = -15$ **(b)** Translation $\begin{bmatrix} 0 \\ -16 \end{bmatrix}$

(c)   **(d) (i)** −41.5   **(ii)** 11.5

(e) 4.087